BUZZKILL

A WILD WANDER THROUGH THE WEIRD AND THREATENED WORLD OF BUGS

BRENNA MALONEY

Illustrated by Dave Mottram

Godwin Books
Henry Holt and Company
New York

Henry Holt and Company, *Publishers since 1866*
Henry Holt® is a registered trademark of Macmillan Publishing Group, LLC
120 Broadway, New York, NY 10271 • mackids.com

Our books may be purchased in bulk for promotional, educational, or business use. Please contact your local bookseller or the Macmillan Corporate and Premium Sales Department at (800) 221-7945 ext. 5442 or by email at MacmillanSpecialMarkets@macmillan.com.

Library of Congress Cataloging-in-Publication Data is available.

First edition, 2022
Book design by Sarah Nichole Kaufman
Printed in the United States of America by Lakeside Book Company, Harrisonburg, Virginia

ISBN 978-1-250-80103-6
10 9 8 7 6 5 4 3 2 1

This book is dedicated to Steve Levingston,
who told me I *could*,
then told me I *should*.

But I wouldn't have, Steve.
Not without your support.

CONTENTS

1.

SCOPE AND SCALE

A COCKROACH CAN live without its head for 168 hours. Moths can't fly during an earthquake. The larva of a Goliath beetle can weigh as much as a McDonald's Quarter Pounder.

How much of that did you know?

On any given night, you could have up to 10 million dust mites in your bed. The praying mantis is the only animal on Earth with one ear. Bees can recognize human faces. Aphids are born pregnant.

I could go on all day. Really. The insect world is a fascinating place. You might not think so. You might think insects are gross. Or scary. Or dangerous. They get a bad rap, insects. Surveys tell us that 25% of Americans are *afraid* of insects. They bite. They sting.

They carry disease. They ruin our picnics. They look icky. If I asked you to name three icky bugs right now, I'm sure you could do it. You might rattle off spiders, ticks, and centipedes. All super icky, I agree! Except that none of those things are insects.

Huh? Yeah, no. None of those things are insects. Spiders and ticks are arachnids. A centipede is a myriapod. So you can go right on thinking those things are icky. I'm right there with you. Yuck-o! But *insects*—insects are surprisingly cool. They have many admirable, enviable traits. They have superpowers. And they are among the most successful living things on the planet.

Maybe you've never really thought about it before. That's okay. You're busy. We're all busy. It's not top of mind; I get that. But I'd like to tell you some amazing things about insects, because this stuff is worth knowing. I'll explain why a little later on, but first I want to tell you what an insect *is* and what it *isn't*. Once I do that, you'll be able to easily pass the spider-tick-centipede test.

Ready? Okay. All adult insects have three body segments: a head, a thorax, and an abdomen. So right from the start, you know a spider isn't an insect. It's only got that tiny little pinhead up front and that big booty in the back. Two parts, not three.

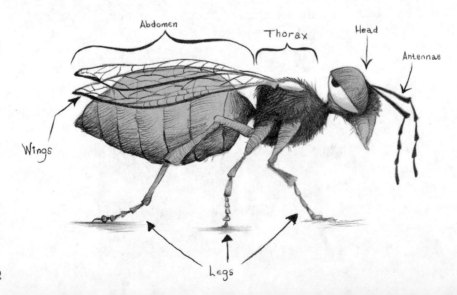

Insects are the only animals in the world that have six legs. Count them up—spiders? Eight. Ticks? Eight. Centipedes? Ha ha, that's a trick question. I want you to answer "100" because *centi-*, in units of measurement, means "hundred." In actuality, centipedes have one pair of legs for each body segment. Depending on the species of centipede we're talking about, they can have between 15 and 177 pairs of legs. Which is way more than six, anyway. So centipedes are *definitely* not insects.

What else? Most insects have wings—two pairs is standard, some (flies) have only one pair. Besides birds and bats, insects are the only animals that can fly.[1] No wings on our spider, tick, or centipede friends.

Insects usually have two feelers or antennae on their heads that they use for various things, such as smelling, listening, or tasting. Insects are also spineless. Every last one of them. I'm not suggesting they lack courage—I mean it in the literal sense: Insects don't have spines. They're invertebrates. You probably know some other invertebrates like octopuses and jellyfish. But insects aren't squishy and shapeless like those guys because they have skeletons on the outside of their bodies called exoskeletons.

So those are the basics.[2] Now you can look at any gross little thing, and without really knowing what it is, you can definitely say whether or not it's an insect. Dragonfly? Three body segments, wings, six legs, antennae. Check! Wood louse? Hmmm . . . Sounds insect-y. But no! Once you take a look at it, you can see it's only got one main body segment. The wood louse is actually a

---

[1] Flying squirrels don't really fly. You know that, right? What they do is more of a *gliding* action. Same for Wallace's flying frog. Just saying.
[2] There are exceptions to every rule. If you look hard enough, you'll find them: The cave midge lacks eyes, silverfish lack wings, protura lack antennae, etc.

crustacean like a crab or a shrimp. Slug? C'mon! That's too easy: no legs, no wings, no exoskeleton. It's just got that one blobby body with doodly-doos on its head. Definitely not an insect.[3] See? Once you know the rules, you can get pretty good at this game.

Earlier I told you that insects are some of the most successful animals on the planet, and I want to circle back to that idea and tell you more. How do we measure success? One way is to look at the sheer volume of insects on our planet.

BY THE POUND

You probably learned this in the fourth grade, but it might be lodged deep in your gray matter . . . somewhere between that list of state capitals and that other list of American presidents. No worries. I can remind you. The animal kingdom is made up of six categories: mammals, amphibians, reptiles, birds, fish, and invertebrates. Invertebrates make up the largest chunk—more than 95% of the animal kingdom. Insects make up 71.1% of the invertebrate group.

Close to a million *species* of insects have been described and named. These are the ones we know about. Scientists have only been able to come up with rough estimates of the total number of insect species on Earth. They put that best guess somewhere between 3 million and 100 million species.

What does that mean in terms of the actual number of insects? It's like a crap-ton. Obviously, "crap-ton" is not an objective number. I merely say that so you will know there are a lot. Some

---

[3] It's a shell-less mollusk, if you must know. And those head doodly-doos are actually called tentacles. Most slugs have two pairs. The upper pair is light sensing and has eyespots at the ends, while the lower pair provides the sense of smell. Which is all very cool. But. Not. An. Insect.

people have trouble processing big numbers. If it's more than, say, five, I have trouble. But for the sake of transparency, I will give you the current estimate: At any time, it's estimated that there are some 10 quintillion individual insects milling about.

What even *is* a quintillion? I had to look it up. It's a 10 with 18 zeros after it. (Like that helps.) Here are a couple of other, more digestible ways to look at it. One out of every four animals on Earth is a type of beetle. For every human on Earth, there may be more than 450 kilograms (1,000 pounds) of termites—about as much as a full-grown cow. There are 1.4 *billion* insects *per person* on this planet. Like I was saying: crap-ton.

BY THE PLACE

You also probably know this already, but insects are *everywhere*. They live in every biome and every habitat on Earth. You can find them in wetlands, tundra, grasslands, oceans, deserts, tide pools, mountains, forests, prairies, rain forests, caves, swamps, in people's homes, on people's bodies, and in Antarctica. Antarctica always gets dissed. Like *nothing* goes on there or something. Tell that to *Belgica antarctica*, baby. It's a flightless midge. It's pretty hard to see as it's no bigger than 6 millimeters (0.24 inches) long. Still, it's the largest purely terrestrial animal native to the continent (as well as its only insect). Most animals can't hack it there—it's too freaking cold. Temperatures there cause cells to freeze and expand, and they can cause irreparable damage to most living things. Not so the midge! It's capable of dehydrating its tissues to survive in temperatures down to 15° Celsius (5° Fahrenheit).

Belgica antarctica isn't the only insect that can survive in extreme temperatures. Consider the Sahara Desert ant (*Cataglyphis bicolor*).

This little dude scoots out of its burrow at the hottest part of the day, when the surface temperature of the sand is hitting about 70°C (158°F). What better time to dodge predators that can't take the heat? The desert ant is looking to feast on the corpses of insects that have succumbed to the elements. How are these ants active in temperatures that would kill most living things? The desert ant has three tricks for its survival. First, it's fast. It can hustle at 1 meter per second. Second, its long legs lift its body to a height that's actually 6–7 degrees cooler than the ground. And third, it only stops moving when it's in the shade. The Sahara Desert ant is bad to the bone.

But these tough little buggers don't hold the record for heat. You've got your fire beetle to consider for that. While the desert ant lives in the desert and can't help being around the heat, the fire beetle *seeks* the heat. It will fly 129 kilometers (80 miles) out of its way to cross paths with a raging forest fire.[4] Why? Because these wood-boring beetles mate while a forest is still burning. Females then tuck their eggs under smoldering tree bark. The eggs hatch into larvae that nibble on what's left of a tree for up to a year before tunneling their way out as adult beetles. It's pretty clever, really. If they tried to breed in living trees, they'd be goners. The tree's cell growth would squash them, or its resin would smother them.

You'll find insects in high places. Some butterflies have been observed flying at altitudes up to 6,000 meters (almost 20,000 feet). Many insects make their home in the soil. You'll find blind, cave-dwelling insects deep underground.

So we've covered cold places, hot places, high places, and low places. I think you get the picture: Insects are in a lot of places. But indulge me a bit further while I belabor the point. I need you to

---

[4] How? Their bodies can detect infrared radiation.

understand that insects really are *everywhere*. In California, there's a maggot that lives in puddles of crude oil. It's called the petroleum fly. I'm not making that up. *Helaeomyia petrolei*. It can submerge itself completely and swallow oil with no ill effects. It feeds on other unfortunate insects that fall into the puddles. It's the only known insect species that develops in crude oil, a substance that is normally toxic to insects and, really, to every other living thing.

So, you know, that's weird. But not nearly as weird as *Orthohalarachne attenuata*. That's a type of mite that lives in a walrus's nose. Don't look it up. I'm warning you right now; it's icky.

You looked it up, didn't you? Well for Pete's sake, what did you expect? I *told* you where it lived. Shouldn't that have been enough for you? Did you read up about them too? Did you see the part where they can be sneezed up from one walrus to another? Good grief. And did you read about that one guy? At SeaWorld? How he saw the walruses spitting and snorting and then later reported an irritation in his eye? Yeah, you know what they found, don't you?

And speaking of eyes . . . you know that they live in your eyelash roots, right? Not *Orthohalarachne attenuata* but *Demodex folliculorum* and *Demodex brevis*. Both of these types of mites are microscopic, so you can't see them with the naked eye (especially when they are *in* your eye). They hang out there in your lashes and feast on dead skin cells because, you know, yum, yum!

I'll give you one more example, just to make my point crystal clear. I'm guessing you've never really given any thought to moths and their ears, much less to moths and their great difficulty with ear mites. It's an issue. *Dicrocheles phalaenodectes*.[5] These mites break through the tympanic membrane of a poor moth's ear and set up a

---

[5] You do know that I can't pronounce any of these Latin names, right? I'm just sticking them in here because I know you're going to want to look them up.

small colony there. A moth needs its hearing to avoid predators—namely bats. Yet, once that membrane is broken, the moth goes deaf, which is why the mites only ever attack a moth in one ear. A moth with one functional ear is still able to avoid bats. I'm not sure how the mites instinctively know not to harm the other ear, but they do.

I've got a million more examples. You know I do. I told you: Insects are everywhere. Yet insects aren't just successful because they outnumber us and because they are all over our planet.

They succeed because their bodies are wildly diverse, and over time, their bodies have adapted to many different environments. With people, one body type has to basically fit all of us. We've all got eyes, mouths, noses, and ears, and they all function in about the same way. Within the insect world, you see a great deal of variation.

MIGHTY MOUTH

Let's start with the ol' piehole. People can be bigmouthed or mealymouthed, but people generally have the same anatomical mouth. Insects have different kinds of mouths, and the type of mouth they have determines what and how they eat. If you're an insect, you're a chewer, a sucker, a sipper, or a sponger.

Chewing mouthparts are the most common. Think of grasshoppers and crickets. These insects use one pair of jaws (mandibles) to masticate. Such a great word, *masticate*. It means to bite, cut, tear, crush, and chew food. These jaws move side to side. If you're a meat-eater, your mandibles are probably serrated and knifelike. If you stick to plants, your mandibles are probably broader and flatter. A second pair of jaws (maxillae) helps shove the food down your throat.

If you're not a chewer, you might be a sucker. The best example of this, of course, is the mosquito. A mosquito's piercing, sucking mouthparts work essentially like a hypodermic needle. Mosquitoes poke a hole in their victim and suck blood through the same opening. Insects like aphids and stink bugs aren't interested in blood; they only want to suck plant juices. But their mouths work in the same way.

Suckers sometimes have special spit that's loaded with digestive enzymes. It breaks down food for easier sucking. Enter the assassin bug. It impales then injects its prey with a venom-laced saliva. The saliva paralyzes the prey and liquefies its internal organs. Assassins are so efficient with their technique, they're able to suck out more than 90% of the live weight of their victims.

A more polite way to eat is to sip your food. The siphoning mouthpart is a friendlier version of the piercing and sucking syringe. It's the long, flexible straw, or proboscis, that butterflies and moths have. When not in use, the proboscis is curled up tightly beneath a butterfly's head. The butterfly unfurls it to sip flower nectar.

A fly can't chew, suck, or sip food. Instead, its mouthparts are equipped with a sponge. To make food soft enough to mop up, a fly has to barf on it first. That's right. Flies puke up an enzyme-laden saliva that quickly dissolves food. Then the fly laps that mess up—barf and all—with its spongy mouthparts. Flies aren't picky eaters, but they do like to taste what they're feasting on first. They do that with special receptors in their feet.[6] A fly's feet are, in fact, about 10 million times more sensitive than a person's tongue.

What's interesting about these different mouthparts is that some insects have more than one type during their lives. As larvae,

---

[6] Butterflies can do this too.

caterpillars are chewers, but when they become adults (butter-flies), they develop proboscises and become sippers. Still others have mouthparts that are a combination of the types I've described. Bees have chewing mandibles, but they also suck liquid through a beak-like tongue. A few insects, like mayflies, have no functional mouthparts at all. They live only a few days with the sole purpose of mating and laying eggs, so they don't need to eat.

CLEAN YOUR PLATE

What do insects eat? Whatever they want! Lots of insects eat plants, some insects eat other insects, and some even drink blood. As I mentioned, mayflies and some moths never eat. Their life cycles are too short for them to bother. Other insects—I'm looking at you, silkworm—cannot seem to step away from the table. A silkworm eats enough leaves to increase its weight more than 4,000 times in just 56 days.

The bulk of insects are herbivores. Some are picky eaters. The caterpillar of the monarch butterfly, for example, eats only milk-weed leaves. It will starve to death if it cannot get these leaves, and once it has them, it's capable of buzz sawing its way through an entire leaf in under four minutes.

Termites are destructive wood-eaters. They get into damp wood through the ground and bore elaborate tunnels as they dine. On its own, a single termite can't do too much damage, but since they typically live in colonies of 50,000, they can be formidable. The largest subterranean termite colonies eat about 0.45 kilograms (a pound) of wood per day.

About a third of all insects are carnivores. I'm sure you can think of many examples of meat-eating predator insects: dragonflies, praying mantises, mosquitoes. I think carnivores stick in our minds more than herbivores because they're so dramatic. They don't just nibble their salads quietly while reading a good book in the corner. No, they tear heads off and let the juices run down their chins.

It's no surprise praying mantises are killers. They *look* lethal. But a sweet-looking harlequin ladybug (*Harmonia axyridis*) may eat as many as 5,000 aphids during its life cycle. And not all insect prey is small. Giant water bugs (*Lethocerus americanus*), which may be as long as your hand, can kill tadpoles, minnows, and small frogs for food.

And I don't mean to scare you, but driver ants (*Dorylus laevigatus*) go after extremely large prey. They are the only insect known to attack and devour *people*. You heard me. They will form columns 50 million strong and strike off across the countryside in search of food. If you are in the way, they will take you out. Now, to avoid causing a panic, I should say this happens rarely. But geez. It *happens*, people. It happens.

Still other insects, like the cockroach, will eat almost anything. These are omnivores. Cockroaches will dine on anything in your home: candy, cheese, meat, grease, fruits, vegetables, dog food. They'll also consume leather, beer, glue, dried skin, books, paper, and human dander. The list is endless.

Ants are also omnivores. They tend to stick to living and dead plant matter, but they are foragers. A single ant can consume up to one-third of its body weight at a time. Yet each ant also contributes to the food supply of the entire colony. While one ant may only eat ounces per day, the entire colony might harvest pounds.

A cricket's diet is very similar to a person's diet. They are omnivores that eat fruit, meat, and vegetables. Their outdoor diet is a little different from ours: rotting leaves, rotting fruit, and insects. And they have a strange fondness for fabric.

Silverfish gravitate toward carbohydrates, particularly sugars and starches. They also eat shampoo, glue in books, paper, clothing, flour, and linens. You know, whatever you have on hand and can spare. They aren't picky.

As always with insects, there are many, many stress-inducing examples I could give you of insects eating disturbing things. If you start to look into it, you quickly discover that there's an uncomfortable amount of blood being sucked by insects. *I* thought so, anyway. On average, a female mosquito that's ready to breed can drink between one and three times her own weight in blood.

Unlike mosquitoes, *all* bed bugs need a blood meal to survive. Research shows that bed bugs take between 5 and 10 minutes to fill themselves up. Okay, granted, bed bugs are small, and you might not even notice them feasting on you until well after the fact. But still. C'mon! Super gross.

THE BETTER TO HEAR YOU WITH

Many species of insects have exquisitely sensitive ears.[7] You just rarely find their ears on their heads. Mosquitoes hear through their antennae. Crickets listen through their legs. Lacewings have ears on their wings. Cicadas and grasshoppers pick up sound through their abdomens. Parasitic flies hear through their necks. Among butterflies and moths, ears turn up everywhere, even on mouthparts.

In some species, ears are plentiful. The bladder grasshopper (from the family Pneumoridae) has six pairs of ears that line its abdomen. Yet the praying mantis has only one ear. And it happens to be in the middle of its chest.

---

[7] Except for beetles, which are almost all deaf

This might seem all over the map to you—and rightfully so. But remember, insects are equipped with what they need to survive. We know sound is created by vibrations carried through the air. An insect's ability to hear means it has one or more ways to perceive and interpret those vibrations. Ears capture, amplify, and filter sound. They keep track of predators, prey, and mates. And just as with insect mouths, insect ears come in a variety of designs.

Many hearing insects have a pair of organs that operate the way a drum does. A tight membrane stretches over an air-filled cavity. The membrane vibrates when it comes in contact with a sound wave. These tympanic organs are paired with a special organ that translates the sound into a nerve impulse. Butterflies and grasshoppers have these.

Some insects rely on a group of sensory cells on their antennae called the Johnston's organ. Fruit flies rely on this to detect wingbeat frequencies of potential mates. Hawk moths rely on it to help stabilize their flight. A lot of caterpillars rely on small, stiff hairs called setae to sense sound vibrations.

Certain hawk moths have a remarkable structure called a labral pilifer in their mouths that enables them to hear ultrasonic sounds. This tiny, hairlike organ is believed to sense vibrations at certain frequencies such as those produced by echolocating bats. I have this image in my mind of the moth flying around with its mouth open, which might be useful to listen for bats as well as hoovering up any mosquitoes that get in the way of its flight path.

Not surprisingly, the type of ear an insect has relates to the quality of its hearing. Mosquito ears can hear specific frequencies as far away as 10 meters (32 feet); the many-eared bladder grasshopper can hear a kilometer or more away. Cricket ears pick up low frequencies; mantis and moth ears detect ultrasound. Katydids have broadband hearing.

ARE YOU HEARING THIS?

While we're on the subject of hearing, can we talk more about sound for a minute? How do insects make sounds, and who is the loudest? I always thought cicadas held the record for being the loudest, but I've recently discovered that's not quite right.

Insects have a number of ways they make sounds and, of course, different reasons for doing so. They might make a sound to scare off an attacker. They might make a sound to communicate with others of their own species—a mating call, for example, or a threat against an encroaching, competing male.

The most familiar sound-making method is called stridulation. Essentially, this is rubbing one body part against another. This is how grasshoppers, crickets, and some beetles make noise. Longhorn beetles scrape ridges on their heads against their bodies. It sounds a bit like rubbing two pieces of Styrofoam together. Grasshoppers have a series of small bumps on the inside of their hind legs. They rub the bumps against their front wings to make sound. With crickets, the bottom of the left wing is covered with toothlike ridges that make it rough. It's known as the file. The top part of the right wing is called the scraper. When the cricket rubs its wings together, the file drags against the scraper and makes a chirping sound.

Only male crickets chirp, and they do so when they are looking for females. But the cricket's chirp has long been admired by people as a pleasing sound. And it's oddly useful. It can predict the temperature. No kidding. It's a phenomenon known as Dolbear's law. Good old Amos Dolbear, an American physicist and inventor, figured it out in 1897. Here's the formula to convert cricket chirps to degrees Fahrenheit: Count the number of chirps in 14 seconds. Add 40. That's your temperature.[8] I couldn't make that up.

But I digress. Katydids rub their wings together to make a call that sounds like "Katy did, Katy didn't, Katy did, Katy didn't." Not sure what Katy is being accused of here, but it must be something pretty serious.

Cicadas[9] usually get blamed for everything, but remember, they are active during the day. If you are hearing loud insect noises at night, you're probably hearing katydids and crickets.

When they do "speak," cicadas use the tymbal in the front part of their abdomens. It has a series of ribs that buckle one after the other when the cicada flexes its muscles. Every time a rib buckles, the rib produces a click. It's sort of like how you make sound with a bendy straw—pushing and pulling the ribs of the straw together to make that click. If you could push, pull, push, pull, push, pull a bendy straw hundreds of times a second, the clicks would be so close together, it would sound like a buzz. That's what's happening with our friend the cicada. Much of the rest of their abdomen is hollow, so it acts as an amplifier for the sound.

Insects can also make sound by tapping some body part against something else. The deathwatch beetle (*Xestobium rufovillosum*) bangs its head on the wooden rafters of old buildings.[10]

---

[8] To convert to Celsius, count the number of chirps in 25 seconds. Divide by 3 then add 4. That's the temperature in Celsius.

[9] Later in this book, we'll have a field day with cicadas. There are somewhere in the neighborhood of 3,000 species of cicadas—some with really cool names, like the dog-day cicada and the scissor-grinder cicada. Their life cycle is super bizarro!

[10] Superstition has it that a deathwatch beetle's sounds are an omen of impending death.

Flies and bees buzz, using their flight muscles and wings to make the sound. You wouldn't think they were *trying to*, but they actually *are* communicating when they do that. Any beekeeper who's opened their hive and heard the "angry buzz" can attest to this.

And here's something you might not know: Bees make a whooping noise when they are surprised. Honey bees can make a vibrational pulse with their wing muscles that is inaudible to humans but can be detected by accelerometers embedded in a hive. It took researchers a long time to figure out the meaning of this sound. At first they thought it was a way to vocalize "Stop!" Then they thought it might be a way to indicate "Food!" Researchers found that the sound often happens when a bee bumps into another bee near the accelerometer. They now believe that bees make that sound when they are surprised!

Another means of sound production is forcibly ejecting air or fluid from the body. Bombardier beetles make a popping sound when they eject a hot fluid from their backsides. The chemical, designed to ward off predators, vaporizes when it meets air.

Making sound by expelling air is said to be somewhat rare in insects. Scientists were stumped for years by the sounds walnut sphinx moth caterpillars made—loud squeaks and whistles. Caterpillars have eight pairs of openings (called spiracles) on their sides as part of their respiratory system. To discover which spiracles were causing the sounds, researchers systematically blocked each pair in turn. The enlarged openings of the eighth pair were responsible for the whistling sounds. The caterpillar forces the air out by contracting the front part of its body. This is a quick and easy way to make sound when threatened.

Yellow warblers and other foraging birds get spooked by the whistling and typically move away. In some instances, the caterpillar whistles mimic the alarm calls of birds like chickadees. So their whistles kept these birds away too. Madagascar hissing cockroaches use the same technique when they hiss.

This is all well and good, but I know you're wondering the answer to my earlier question: Which insect is the loudest? It's not the male cicada, as it turns out. Don't get me wrong. When male cicadas get together to woo the ladies, the din is impressive. A choir of lovesick cicadas can out-voice a lawn mower at full bore, reaching volumes greater than 100 decibels.

No, the loudest insect is also the loudest animal on Earth. It's the male water boatman. The water boatman (*Micronecta scholtzi*) is a tiny insect that lives at the bottom of freshwater ponds. It's less than half a centimeter long but makes a noise of 99.2 decibels.[11] By itself. For comparison, a freight train is about 90 decibels. *How* he makes this sound is somewhat unusual. He uses his penis.

The water boatman strikes his penis along the ridges on his abdomen. And despite the organ's size—only the width of a human hair—this courtship song renders him the loudest animal on Earth, relative to his size. The sound doesn't carry well from water to air. In fact, at least 99% of the sound is lost. Which is probably why you and I have never heard it.

---

[11] There are animals that make louder calls. The record goes to the sperm whale, which can create clicks equivalent to 170 decibels on land. Other animals, including elephants, hippos, and dolphins can produce louder calls than the water boatman. But pound for pound, there is no competition.

I SPY

Insects and I have one thing in common: We're nearsighted. They can't see an object clearly if it's more than 1 meter (3 feet) away. I can't find my way out of a room without my contact lenses or glasses. Unlike insects, however, I don't have compound eyes, which is just as well because that would be creepy.

Adult insects usually have two compound eyes.[12] Each is made up of many individual lenses[13] called facets. The eyes of some insects have as few as nine facets each; house fly eyes have 4,000 facets; and some dragonfly eyes have nearly 30,000.

Behind each facet there is a tubelike structure known as the ommatidium, which is a fantastic word to know and say. *Ommatidium. Ommatidium.* It's very musical, really. Each one of these ommatidia, or tubes, sends the image captured by the facet to the brain. The brain combines all the snippets together and produces a kind of pixelated image. The more facets there are in the eye, the sharper the complete image.

You might have noticed that a lot of insects seem to have disproportionately large compound eyes, and they are positioned on the sides of their heads. If insects were the size of people, which would cause me to have a minor stroke, their eyes would be the size of footballs. The size and position of their eyes make it possible for them to see in most directions.

While compound eyes are the main organs of sight, most adult insects have another set of eyes too. They have two or three simple eyes, each with one lens. They are called ocelli. The ocelli are

---

[12] Insects don't have eyelids, and honey bee eyes are hairy. I just need to blurt stuff like that out. It's hard to carry all this information around in my head without it bubbling up. I lose friends this way; I really do.

[13] A person's eye has only one lens.

located at the top of the insect's head and sometimes form a little triangle there.

Unlike the rest of us, insects cannot turn, move, or focus either set of eyes. Simple eyes help insects detect light and dark; compound eyes help insects see the world in greater detail—and they are especially helpful in detecting movement. Movement is key for many insects. Predatory insects cannot see their prey unless it moves, even if it is right in front of them.

Insects can see some colors, but their view of colors is different from ours because they see on a different color spectrum than we do. They can see ultraviolet light. Bees use ultraviolet light to find flowers that have lots of pollen in them. Other insects, like butterflies, use ultraviolet light to find a mate.

So insect eyes are usually found on their head, and they usually have both simple and compound eyes. Usually. I mean, there are exceptions. And you probably figured out by now that I feel some sort of obligation to tell you about the exceptions, right? Because if I don't tell you, then I have to keep these things to myself, and somehow, that would be wrong.

That said: A male yellow swallowtail butterfly sees with his penis. You heard me. His penis is lined with photoreceptor cells, which help him see. Apparently, the ability to see light is important to ensure his success. Yellow swallowtails mate facing away from each other. The males use their light-sensitive genitalia to line up with females properly.[14]

Having a variety of eyes, ears, and mouths helps insects succeed in a lot of different places. What we'll now see is that, as a group,

---

[14] It was Kentaro Arikawa, professor at the Graduate University for Advanced Studies in the Department of Evolutionary Studies of Biosystems in Japan, who made this discovery. His academic interest is in how insects see in color.

insects also share certain universal qualities that help them to succeed. Insects are armored; they're small; they aren't picky eaters; they fly; they metamorphize; and they have great strategic defenses.

FROM THE OUTSIDE IN

It might seem weird to have your insides on your outsides, but an exoskeleton is a must-have for insects. It's like the ultimate power suit. Not only does it look good but it also performs a number of key functions: It protects internal organs, prevents dehydration, attaches to muscles, and allows an insect to gather information about its environment.

European knights during the Middle Ages had their own version of the power suit—armor worn into battle. A full suit had a helmet, a breastplate, gauntlets, a gorget (to protect the neck), cubitieres (to cover the elbows), underarm protectors (not the deodorant kind; more like star-shaped, metal things), cuisses (to cover the thighs), greaves (to guard the ankles), and sabbatons (to blanket the feet). They also had the option of wearing chain mail, a type of flexible armor made of iron rings linked together. It came in the form of hoods or long shirts, which could give some protection to a knight's nether regions.

An insect's exoskeleton also works to protect. It's built to resist both physical and chemical attacks. It's similar in design to the suits of armor because it's also made up of curved plates and tubes that link together. The underbelly of the exoskeleton is probably the softest part, but the backs of insects are often covered with thick, reinforced plates. Legs are hollow tubes. All parts of the insect are protected—even the eyes, which are covered with a transparent exoskeleton.

I know I told you that all insects have exoskeletons, so you probably want to challenge me on that. *What about caterpillars? Caterpillars are squishy.* They are; it's true. But even caterpillars have a thin exoskeleton. Caterpillars keep their shape because they are under pressure. They're like living balloons. Their body fluids press outward against their thin exoskeleton, stretching it and keeping it tight. While their exoskeletons are thin and "soft," many insects find themselves on the other end of the spectrum. The ironclad beetle (*Zopherus nodulosus*) has one of the hardest exoskeletons in the animal kingdom. You can literally stomp on it, and it will take zero damage. You would need a power tool to cut through its exoskeleton.

A full suit of medieval armor weighs a lot—about as much as your average seven-year-old. Imagine carrying one of those on your back as you head into battle. But remember, an insect has to be able to fly with this thing on, so it needs to be lightweight. An exoskeleton is made up of proteins and chitin, which sounds like a key ingredient in many Southern dishes. I looked it up, and one definition was "a long-chain polymer of N-acetylglucosamine." I didn't understand that, of course, so I tried again. Another definition was "a carbohydrate substance." I can think of lots of carbohydrate substances. Isn't that what Pringles are made out of? Probably not the same recipe, though.

The outer layer of exoskeletons is covered by a thin layer of wax. This impermeable barrier makes insects water-repellent. The wax also traps moisture in, so insects don't dry out.

That said, the exoskeleton can be strangely porous. Depending on the species, insects can detect light, pressure, sound, temperature, wind, and even odor through their exoskeletons.

Freedom of movement is ensured by membranes and joints in the exoskeleton. Muscles that attach directly to the body wall combine maximum strength with near-optimum leverage. Have you ever wondered how insects can be so strong? The dung beetle (*Onthophagus taurus*) can lift weights up to 1,141 times its own body weight—a load equivalent to a human lifting more than 81,000 kilograms (about 180,000 pounds). One reason for their strength is the thickness of their muscles. Muscle strength depends on thickness, not length. Another reason for their strength is that they have more muscles than many animals. A person has about 650 muscles. Some caterpillars have about 4,000.

Of course, the internal anatomy of insects is very different from ours. Theirs is an open circulatory system, which means that their blood sort of sloshes around inside their exoskeleton. Insect blood doesn't carry oxygen like ours.[15] Instead, the long tube that makes up an insect's heart pumps blood onto the brain, which then flows backward through the body, bathing the organs, muscles, and nervous system. It brings digested food and takes away waste materials.

You'll find the brain in the head, although some information is also processed at "mini-brain" nerve centers elsewhere in the body. Which is why you'll sometimes see a disembodied leg continue to kick or a headless body running around, seemingly unfazed.

You and I use our lungs to inhale oxygen and exhale carbon dioxide. Insects don't have lungs. Instead, they take in oxygen

---

[15] Insect blood is greenish or yellowish in color because it lacks hemoglobin, which is what makes our blood red.

through tubes called tracheae. Breathing holes called spiracles, which I mentioned earlier, connect to the tracheae, allowing air in and out.

The inner and outer workings of an exoskeleton are nothing short of miraculous. Here's the bummer, though: As awesome as the exoskeleton is, it can't grow. It's not living tissue. In order to get bigger, an insect has to molt and shed its exoskeleton.

This is something that every kid learns about in maybe the second grade. I never really gave it much thought before, but let me tell you something: Molting sucks. Not only is it darned uncomfortable but it's also dangerous. It leaves the insect extremely vulnerable for a period of time.

The new skin an insect needs has to be bigger than the old skin it currently has. And the new skin begins forming under the old one well before molting takes place. It's a little like wearing an overcoat *under* a sweater for a while.

The whole process is triggered by dramatic hormonal changes in the insect. When the hormones fully spike, the insect experiences strong muscular contractions, and the insect gulps in air to cause its body to swell until the old skin splits open.

The next part is tricky because the insect has to shrug off the old skin. Picture trying to peel off a sweaty T-shirt after you've been outside on a hot day.[16] Woe be to the insect that gets stuck or accidentally rips off a leg because it gets tangled up.

Once free from the old skin, the insect is vulnerable. It literally cannot move until the new skin hardens, which could take several hours. Some estimates put insect mortality as high as 85% during molts.

---

[16] Remember that the exoskeleton covers the entire body, including the eyes. I keep reminding you about the eye part because it skeeves me out so much.

Molting is a lot of trouble to go through, but it's mandatory. Some insects hold the record for molts. The firebrat (*Thermobia domestica*), a small insect similar to a silverfish, molts 60 times. But a species of mealybugs (family Pseudococcidae) only has to go through this nightmare twice in its lifetime.

SIZE MATTERS

Generally speaking, insects are itty-bitty. To someone who is squeamish, the biggest of them can get uncomfortably big. The South American longhorn beetle (*Titanus giganteus*) can grow to be 16.7 centimeters (6.5 inches) long, not counting its antennae. The giant wētā of New Zealand is probably the heaviest. I saw a picture of one once next to a mouse. *They were the same size.* One of the longest, of course, is Chan's megastick (*Phobaeticus chani*) at 62.4 centimeters (24.5 inches).

But the smallest insect could fit inside the period at the end of this sentence. I'm going to say *Dicopomorpha echmepterygis* is the smallest. This parasitic wasp is a miraculous creation. Its tiny body—neatly packaged with complete digestive, reproductive, respiratory, and circulatory systems—is actually smaller than a single-celled amoeba. These little insects are known to hitchhike on the faces of butterflies and larger insects to get around.

Why does size matter? For an animal with an exoskeleton, smaller is better. Think about it: If an insect was the size of an elephant, frankly, I would pass out. No, that's not what I was going to say. If an insect was as large as an elephant, its exoskeleton would have to be massive to support the additional body mass. The simplest movements would be exhausting to the poor thing as it tried to haul itself around. There is an upper limit on how large an insect can actually become.[17] Beyond this size, the insect's surface area is just too small for attachment of all the additional muscle tissue.

Another advantage to being small is that you don't need much to survive. You can escape predation fairly easily. If you're being chased, you can squeeze into cracks or duck under leaves. You can hide. You also don't need a lot to eat. Some insects, like the leafminer, spend their entire larval stage tunneling between the paper-thin layers of cells in a single leaf.

WINGING IT

Insects are the only invertebrates that can fly. And flying is everything. It not only gives insects a quick way to escape predators but it also allows populations to quickly reach new habitats and exploit new resources.

---

[17] Since a twofold increase in body length typically results in a fourfold increase in surface area and an eightfold increase in volume and mass, an insect can't get much bigger than 150 grams (5.3 ounces). And it's because of this dizzying math and science that you can be reasonably certain that a creature such as Mothra could never exist. Which is a great relief.

Most insects rely on two pairs of wings, which join or overlap so they work together as a single pair.[18] Some insects, such as flies and mosquitoes, have only one pair. The hind wings have evolved into stumps that help these insects balance while flying. Insects such as fleas, lice, bed bugs, and silverfish have no wings.

Insect wings are thin, like cellophane—one of nature's lightest structures. A network of veins stiffens them for flight. There's a great deal of variation. Butterfly and moth wings are covered with scales. Beetles have a set of hard, shell-like wings called elytra. They fit like shields over their hind wings.

Wings are curved on top and flat on the bottom. Air rushing over the wing has to travel farther because of the curvature, so this air moves faster than air below the wing. Since fast-moving air exerts less pressure than slow-moving air, the difference creates suction, called lift. Each downward wing flap creates more lift, propelling the creature up and forward. Flight is powered by muscles attached directly to the wing bases.

Insects are agile flyers. Dragonflies dash, dart, hover, and fly backward or sideways—and they do so at speed. Some species are remarkably swift. Large hawker dragonflies (family Aeshnidae) have been clocked at speeds of 54 kilometers (roughly 30 miles) per hour.[19] Not all insects fly that fast. The average cruising speed of a house fly is about 19 kilometers (12 miles) per hour.

---

[18] Dragonflies can use all four wings independently.
[19] By comparison, the fastest human sprinters run only about 37.58 kilometers (23.35 miles) per hour.

Some insects travel great distances or remain airborne for long periods of time. More than 200 species—including moths, dragonflies, locusts, flies, and beetles—are known to migrate over long distances by air. One migratory locust (*Schistocerca gregaria*) is capable of flying nine hours without stopping. Large swarms occasionally cross the Mediterranean Sea. If I was on the water and saw a horde of locusts coming my way, I'd be confessing all my sins and praying to Jesus to come save me.

ON THE MOVE

Insects have ways to get around other than flying. Insect legs may be adapted for running, jumping, grasping, and swimming.

Tiger beetles are among the fastest runners on Earth, relative to their size. They clock in around 9 kilometers (5.6 miles) per hour. Stop laughing. I know you could probably outrun this guy, but you're not looking at this the right way. Think of a cheetah, right? Arguably, the fastest land animal on the planet, with speeds up to 120 kilometers (75 miles) per hour. The cheetah can cover 23 body lengths in a second. Impressive. Okay, now back to the tiger beetle. When you do the math, it's moving 171 times its own body length. It runs so fast after its prey that it literally goes blind.[20]

The froghopper (*Philaenus spumarius*) can jump higher than any other animal, relative to its size. I know you think it's the flea. Everybody always thinks it's the flea. It's not the flea. Yeah, sure, fleas are formidable jumpers, but I'm telling you, check out the froghopper. It doesn't look like much, but when it jumps, its hind

---

[20] Cole Gilbert, Cornell professor of entomology, made this discovery. The tiger beetle moves so quickly, it can't gather enough photons (illumination into the beetle's eyes) to form an image of its prey. Although it is temporary, it goes blind.

legs generate a force that's 80 times the g-force experienced by astronauts when they blast off into space.

You ever wonder how a fly walks on the ceiling? When most insects walk, they move three legs at a time. These are the front and hind legs on one side and the middle leg on the other. At the next step, the other three legs move. In this way, an insect is always resting solidly on three legs.

Each fly foot has two thick footpads that give it plenty of surface area with which to cling. These adhesive pads come equipped with tiny hairs that have spatula-like tips. The hairs produce a glue-like substance that helps the fly stick. To unstick itself and move forward, each foot also comes with a pair of claws that pry the gooey foot off the ceiling. The fly has a few maneuvers it can try to unstick: pushing, twisting, and peeling its footpads free.

Whirligig beetles (family Gyrinidae) are the fastest swimmers of the insect world.[21] Though they spend most of their time swimming on the surface of the water, these beetles can dive quickly to capture prey or to escape predators. They can reach swimming speeds of up to 44 body lengths per second.

All six legs are needed for their success. They primarily use their back legs to propel them through the water, while their middle legs are used to steer. They use their powerful front legs for grasping prey. Microscopic bristles extend off their middle and hind legs, arranged like teeth on a comb. These hairs play a key role in generating thrust during swimming by making the legs more paddle-like. When it sweeps its legs through the water to swim forward, the tiny hairs prevent water from flowing through.

---

[21] They are also notable for their divided eyes, which are believed to let them see both above and below water.

A leg with its hairs extended will function as if it were a solid paddle, generating more thrust than if the hairs weren't there. Upon the leg's return stroke, the hairs collapse against the leg to reduce the drag.

THE BIG CHANGE

Do you remember when you were in elementary school, and the teacher would talk about metamorphosis? You know, the four stages insects go through during their life cycle[22]—egg, larva, pupa, adult. I remember this really well because I saw a book when I was a kid that showed a monarch butterfly going through all the stages, and the pictures were lined up in such a way that you could compare them to each other and see all the changes the butterfly was going through. The photos were perfect. Pristine. You could see everything so clearly. Look! Its wings are forming. Look! Its colors are developing. It looked like such a marvelous process.

Did you know that there are places online where you can buy caterpillars with the express goal of watching them metamorphize into adults? It's true. Get on Amazon and see for yourself. How fun! How neat! Once I made that discovery, I had to do it. I plunked down my $27.50 for the "Insect Lore Butterfly Growing Kit-With Voucher to Redeem Caterpillars Later." I especially liked that part about redeeming the caterpillars later.

The kit arrived within a few days. I knew the caterpillars weren't in the box (I had to redeem them later), but I was still pretty excited. Inside was a reusable mesh butterfly habitat. When

---

[22] There are four stages in complete metamorphosis: egg, larva, pupa, adult. Some insects, like dragonflies, go through incomplete metamorphosis: egg, nymph, adult.

I pulled on the Velcro clasp, the habitat popped open and sprang into shape. Ready to go! Wow!

Then I used the little card inside the kit to order my LIVE caterpillars online. Then I waited (not too long, really) before a little box arrived on my doorstep. I was very, very excited to receive the box. But I need to be straight with you: What followed was not the idyllic thing from the book I read as a child. Metamorphosis is not like the pictures. I mean, it is, but it isn't. The whole thing is rather . . . um . . . gooey.

Inside my box were five very hungry caterpillars in a lidded cup. I could barely see them. The instructions explicitly told me DO NOT OPEN THE CUP YET, so I surely did not. But the cup was half-filled with some sort of brown, gloppy stuff that had, in transit, gotten smeared all over the walls of the cup. I couldn't see clearly, but there were (probably) five caterpillars milling about in there.

Not sure about that brown stuff. The instructions said it was nutritious food, so okay, we'll go with that. The instructions also advised me not to touch the caterpillars but to observe them. "Watch carefully!" it said. "Your caterpillars will shed their exoskeletons several times as they grow."

Well, I did watch carefully, but I'll confess, I didn't see them do this. They nibbled a bit but really didn't seem all that hungry to me. On the second or third day, they seemed a little bigger, I guess. Hard to tell. By the fourth day, they weren't really doing much. In fact, they weren't really moving. In fact, I assumed they were dead. I didn't check for the next few days because I was honestly a little distressed by the thought of a cup of dead caterpillars. Oh me of little faith! They weren't dead at all. No! When I looked again, the little devils had crawled up to the lid and created their chrysalides.

At this point, I was all done with the mucky brown stuff. I was supposed to gently pry off the sticky lid and prop it up sideways so the chrysalides hung down against it. This I did with all the precision of someone defusing a bomb. These chrysalides were tiny and seemed so fragile. Having given them up for dead days before, I now felt a guilty responsibility for seeing them through their process safely.

I carefully transferred the lid to the habitat, which I had to hang from the ceiling so my cats didn't help themselves to caterpillar treats. Here's where things got disturbing. What was going on inside those chrysalides . . . It wasn't pretty. It wasn't peaceful. It was actually kind of disturbing and violent. Shaking, twitching, stretching. Frankly, I was completely unnerved by them. I wondered: What is going on, really, inside that chrysalis? Let me tell you.

Essentially, the caterpillar's body must digest itself from the inside out. The same juices used to digest food as a caterpillar are now being used to break down its own body. It releases enzymes to dissolve all its tissues. This is hard to conceptualize without feeling queasy, but at this point, the insect might be more liquid than solid. Within this mess, certain highly organized groups of cells known as imaginal discs are at work.[23] These are undifferentiated cells, which means they can become any type of cell.

Once a caterpillar has disintegrated all its tissues except for the imaginal discs, those discs use the protein-rich soup all around them to fuel the rapid cell division required to form the wings, antennae, legs, eyes, and all the other features of an adult butterfly.

The process takes time—about 10 days—and is an astounding

---

[23] Turns out, when a caterpillar is still developing inside its egg, it grows an imaginal disc for each of the adult body parts it will need as a mature butterfly or moth—discs for its eyes, for its wings, for its legs, and so on.

transformation. When the butterfly emerges from the chrysalis, it weighs about half of its caterpillar self, which shows you how much of the caterpillar's body had to be broken down into energy to fuel the transformation.

The thing about metamorphosis that is so powerful is that it allows insects to excel at different things at different stages in their lives. A caterpillar has a mouth and a digestive system that allow it to do lots of eating. This is the growth phase of its life, so this makes sense. The butterfly doesn't eat much; it doesn't need to. Instead, it now has wings to fly to find a mate and to lay its eggs.

Metamorphosis reduces competition among members of the same species. They aren't looking for the same food. They aren't living in the same spaces. They have different predators. For this to be possible, nearly everything must change. The gut changes because the butterfly doesn't need to eat like a hungry caterpillar. The body changes because the butterfly needs wings and a different mouth.

Days went by for my little caterpillars. Then one morning, I looked into the habitat and five stunningly beautiful painted lady butterflies were looking back at me. I almost wept with relief. I had never seen anything more amazing in my life.

The tops of their wings were black and orange with white spots. The bottoms of their wings were black, gray, and brown. I could see the scales on their wings. Their bodies were fuzzy. I could see their long antennae and their large eyes and their magnificent proboscises. Surely no greater creature had ever been born. I felt such pride. They had made it! And so had I!

The next day, I had the great privilege of releasing them to the wild. I opened the habitat outside and each butterfly climbed onto my hand, one by one. I lifted my hand to the sky and took one last look before they flew off. How lucky I was to have witnessed their transformation. It was truly joyous.

ON DEFENSE

Another very important consideration for the success of insects is that they have a large arsenal of defenses to help them ward off, confuse, or vanquish predators. Take a look at all these insect strategies.

BE STINKY

Smelling bad can be a very effective strategy for some insects. The creatively named stink bug (*Halyomorpha halys*) has a pungent odor that some best describe as cilantro. This would be off-putting to me because I dislike cilantro—not because I'm one of those people who think it tastes like soap. I just don't like the taste, no matter how many people tell me it makes salsa better. A good part of the natural world must be in my camp because when a stink bug gets its cilantro smell on—which it emits through glands in its abdomen—the birds and lizards that would normally be so game to gobble it up are instead repelled.

Swallowtail caterpillars also deploy the stink strategy. They have an impressive organ on the tops of their heads called an osmeterium. It looks a bit like a pair of yellow antlers. When the swallowtail is feeling peeved, it waggles these antlers about in the general direction of whatever is bothering it and releases a funky smell.

Can't tell you what it actually smells like because I've never been attacked by a swallowtail caterpillar before. Anecdotal reports have it smelling like parsley or something citrusy. Those two things are nowhere in the same neighborhood, of course, but neither sound all that offensive to me. Still, if you're a predator, it appears to be quite off-putting.

If you're an insect and you really want to unnerve potential predators, you could take a page from the eastern lubber grasshopper's (*Romalea microptera*) book. When this thing feels even remotely threatened, it moves right to DEFCON 1 by producing an evil-smelling, bubbling froth from holes in its thorax. The stench alone sends most threats running. As the sticky bubbles burst, the smell intensifies. If that doesn't work, this hopper takes the full nuclear option. With precision aim, it ralphs up a repulsive droplet of chemical-rich fluid from its mouth. And *then*, it hisses at you. Seriously? The hiss seems over the top to me.

HOSE THEM DOWN

Again, maybe not a strategy that you and I would feel comfortable deploying, but it's extremely effective with some insects. The approach is simple: Ooze or spray irritating substances on a would-be predator. As the predator stops to clean itself off and mutter, "What just happened?" the sprayer makes a clean getaway.

Oozers practice something called reflex bleeding, which sounds super gross and *is* super gross. A common practitioner of this defense is none other than the ladybug (*Coccinella septempunctata*). A spooked ladybug will release that dark hemolymph fluid (insect blood) from its leg joints. Hemolymph smells and tastes bad and will even stain your furniture. Let that be a lesson to you!

Blister beetles do it too, but their ooze contains a—wait for it—blistering agent called cantharidin. A blast of this stuff really hurts and can cause nasty blisters. But even blister beetles aren't as bad as African bombardier beetles. Gadzooks! They are like a beetle bomb. They store explosive chemicals in two separate chambers in their abdomens. If you upset a bombardier beetle, it will force the chemicals together so they react and explode. You'll hear a loud popping sound. Then a searing-hot, toxic spray[24] will shoot out through a swiveling nozzle that the beetle *can control and aim*. If the spray hits a small animal in the face, that animal is a goner—either immediately blinded or killed outright. The bombardier can activate at a speed of 500 pulses per second.

STAB THEM WITH SPINES

It looks innocent enough . . . You see a fuzzy little caterpillar, and you want to pick it up. Don't do it! Those fuzzy hairs are filled with poison. The word you're looking for here is *urticating*. It's tricky to say and even harder to drop into casual conversation—believe me, I've tried. But what it refers to are special, hollow hairs[25] that are attached to a gland that pumps poison into each hair. If you so much as brush up against them, the hairs will break and release the toxin. It feels a bit like having shards of fiberglass

---

[24] The temperature of the mixture is nearly 100°C (212°F).
[25] You see urticating hairs a lot in the plant world, since plants don't have a whole lot of other ways to defend themselves.

embedded in your skin. It's a common defense with caterpillars—some to a greater degree than others. Do not . . . do NOT touch a saddleback caterpillar. Okay? I know it looks cute. I know it looks fuzzy. But you will regret it.

STING THEM

Bees always get blamed for the stinging, but let's be honest: Yellowjackets are the most aggressive of the stinging insects.[26] The three most common types of stinging insects are apids (honey bees and bumble bees), vespids (wasps, hornets, and yellowjackets), and ants (fire ants).

Most of these insects use their stingers like a needle. The stinger breaks the skin and venom is injected. For honey bees, stinging is a suicide mission. Their stingers are barbed on the end, and once the stinger snags the skin, it's hooked. When a bee tries to fly away, the stinger stays in place, tearing the bee's abdomen open. So for the honey bee, it's one and done. A queen bee has slightly different anatomy and can sting you more than once, as can fire ants and wasps.

Getting stung can happen when you least expect it. I should know. I'm still recovering from what my family and I now only refer to as The Incident.

It was late July and hotter than hot. I was in a hurry to be somewhere but at the last minute, I thought I'd better top off my bees' water supply. (If you are a beekeeper, this is one of the least invasive things you can do to a hive. You don't have to open the

---

[26] I'm trying to throw shade on the yellowjackets. Probably the fiercest stinging insect would be the Africanized "killer" bee (a hybrid of *Apis mellifera*). They are well-known for their excessive aggression and defensiveness. If you disturb their nest, they will pursue you great distances—up to 400 meters (a quarter of a mile). And once they catch up to you, expect multiple painful stings.

hive to do it—the water container is on the outside, sort of like what you might hook on the outside of a hamster's cage.) The bee suit is great protection but so, so heavy, and like I said, it was hot that day. I thought I'd just dash out with gloves and a veil and be okay. So armed with only a bee brush to sweep the bees away, I went out, pulled the near-empty water container, refilled it, and using the brush, I gently swept the bees from the entrance so I could replace the full water container. That's when I felt the first sting on the outside of my ankle.

I yelled and jumped back just as a second bee stung me on the inside of my calf, same leg. They teach you this in beekeeping class—that when a bee stings, it releases a pheromone that the other bees can smell. It's a chemical message that tells everyone that you are a threat. It puts them on red alert.

I hobbled away from the hive quickly but not before two more bees got me on the outer leg. By the time I reached the door to the sunroom of my house, I was being trailed by a stream of angry bees; two bees were hanging from my leg, dead or dying, and a fifth bee was coming in for the kill on my inside calf.

I don't mean to be a baby about this, but getting stung hurts. And getting stung by five wrathful bees and being chased by a cloud of maniacal bees was scary. My body kind of went into shock from the pain, the indignance (*How* could *they?!*), and the embarrassment (*I should have known better*).

From an objective observer's standpoint, the whole thing must have looked ridiculous. My husband, Chuck, caught sight of me from the house. He said it looked like I was performing some kind of interpretive dance as I hobbled away from the hive and tried to get to the house—swatting and screaming as I went. Once inside the sunroom, I savagely smacked my own leg to beat back the bees that had followed me in. Chuck stood on the other side of the glass door, afraid to open it lest the bees enter the house. Our eyes met. It was like one of those space movies where the hero is on the wrong side of the airlock.

When I finally got the last bee off me, I cracked the door open from my side and yelled, "I NEED A CREDIT CARD!" as if the whole experience left me desperate to charge something. I had to explain to him later, much later, that it's how you get the stingers out—you flick them with a credit card. The stinger has a mechanism in it where it keeps pumping the toxin in even after the bee has stung you, even after the bee has died, and even after the bee has been knocked free from your body. If you don't get the stinger out, the whole situation gets worse for you later. You need to study the wound, though, and flick in the right direction. If you flick in the wrong direction, you drive the stinger in deeper.

I managed the job through tears of pain and betrayal, then immediately took a healthy dose of aspirin and Benadryl to keep the swelling in check. It was my own fault. Bees in late July are not to be trifled with. I learned my lesson. Of course, my pain didn't measure up to the pain that some people are willing to put themselves through.

Your personal hero in the art and science of stinging should be Justin O. Schmidt, a biologist at the Southwest Biological Institute.[27] He knows more about stinging than anyone else on the planet because of his bizarre and largely self-imposed mission to get stung. Like, a lot. Falling under the category of "seemed like a good idea at the time," Schmidt went on a professional quest to

---

[27] Justin O. Schmidt is the author of *The Sting of the Wild*. It should be required reading for everyone. The man is brilliant.

study the impacts of stinging insects on humans, using himself as the main test subject.

During the course of his long career as a working entomologist, Schmidt has been stung by all manner of insects, including such beasts as the cactus bee, the hairy panther ant, and the four-toothed mason wasp. Over time, he began to create an index, which he later dubbed the Schmidt Pain Index. It compares the pain of various stings. They say pain is subjective. Schmidt measures it on a four-point scale with four being the most painful. He also includes a vivid description of the experience—so crystal clear, you would hardly need (or want) to share his experiences firsthand.

Of the red fire ant (*Solenopsis invicta*)—a level-one pain—he writes that getting stung is "sharp, sudden, mildly alarming. Like walking across a shag carpet and reaching for the light switch." The glorious velvet ant (*Dasymutilla gloriosa*) rates a two on the scale: "instantaneous, like the surprise of being stabbed. Is this what shrapnel feels like?" The red paper wasp (*Polistes canadensis*) was a clear level-three pain, which he described as "caustic and burning, with a distinctly bitter aftertaste. Like spilling a beaker of hydrochloric acid on a paper cut." The bullet ant (*Paraponera clavata*) took the form of a level-four pain: "pure, intense, brilliant pain. Like walking over flaming charcoal with a 3-inch nail embedded in your heel." Clearly, the sting's the thing. Avoid it at all costs, I say.

WEAR A WARNING

It's pretty simple: Certain colors are warnings. In the animal kingdom, this is called aposematic coloration. It means your colors send a message to the greater world. For insects, that message may mean: "I taste bad!" or even "Eating me will kill you." Certain color schemes are big no-nos in the animal world: red and black (think ladybugs and milkweed bugs), black and orange (monarch butterflies), and yellow and black (bees and wasps).

BLEND IN

Blending in! Ah, much more benign than stinging. To be a master of disguise, you need to master something called crypsis. When I first heard this word, I thought it might be a new kind of breakfast cereal. It sounds kind of delicious. Crunchy. Stays crispy in milk. But that's not what it is at all. Crypsis is the ability to avoid detection by blending in with your background. This tactic involves not only matching the colors but also disrupting the outline of the body so that no one notices you're there.

The salt-and-pepper moth (*Utetheisa lotrix*) is brilliant at this. It is white with black speckles across its wings. This patterning helps it camouflage against the lichen-covered tree trunks that it rests on during the day. In the same way, the mottled sand grasshopper (*Spharagemon collare*) is extremely challenging to spot in its habitat of sandy-soiled, grassy areas.

Crypsis is a super, super cool defense. The only problem with it is that you have to stay still. You can't go wandering off because, if you do, you'll no longer match your background. That's why a lot of these dudes tend to move very little during the day, and when they do, it's slow and deliberate to avoid attention. They usually stay close to home or make only short trips out and back because their coloration tends to be area-specific.

You might not know this next word: *mimesis*. It's also a form of blending in, but it's hiding in plain sight by resembling something else. For example, stick insects very much want to come across looking like sticks because predators who eat insects have no interest in actual sticks. So the more realistic looking, the better. Some of them are quite convincing.

The thorn bug (*Umbonia crassicornis*) is rather proficient at this ruse too. If enough of those buggers line up on a branch, predators just run away screaming—no way they want to get stuck by all those prickly thorns. And then, ha ha! The thorn bug has the last laugh. See how that works?

As a caterpillar, the brimstone moth (*Opisthograptis luteolata*) looks shockingly like a twig. *Uropyia meticulodina*, a small moth capable of mimicking a dead, curled-up leaf is also pretty amazing. But if you really want to see the master, look no further than the giant Malaysian leaf insect (*Phyllium giganteum*). Really, it's just spectacular.

Some insects use mimesis in a different way. They piggyback on a good thing that other insects developed the hard way. For example, no one bothers monarch butterflies (*Danaus plexippus*) because they taste really, really bad, and everyone knows that because they sport the black-and-orange warning-color combo. Now it just so happens that viceroy butterflies (*Limenitis archippus*) are black and orange too. The two species look suspiciously similar—same general shape, same poisonous colors. Except the monarch is legitimately poisonous (because it chows down on poisonous milkweed all the time and the chemicals get transferred), but the viceroy is not poisonous. In fact, it's probably pretty tasty. (To a predator. Not to me. I don't eat butterflies.) This is plagiarism at its finest, folks! Robber flies do it too by mimicking wasps.

DISGUISE YOURSELF AS SOMETHING SCARY OR DISGUSTING

Many butterfly and moth species have developed what appear to be extra eyes. These distinct wing markings are often referred to as eyespots or false eyes. The idea behind them is that predators will be frightened of the big eyes, and they won't attack. For some species, the spots can't prevent an attack entirely, but they can distract and divert a predator. The squinting bush brown butterfly (*Bicyclus anynana*) has a series of eyespots along the outer edges of its wings. If a predator bites the wings (and not the critter's head), it has a better chance of survival.

If you want to see something really strange (and fear-inducing), check out the snake mimic caterpillar (*Hemeroplanes triptolemus*). Not only does it have false eyes but it also has the ability to expand the front segments of its body. It looks . . . it *really* looks like a snake.

Many caterpillars have a small horn at their back end. When threatened, the gaudy sphinx moth (*Eumorpha labruscae*) will lower its horn to look like a snake's flicking tongue. Combine that with its eye spots, and you've got a pretty convincing mimic.

Looking like something nasty is also a successful strategy. You can't go wrong with poop. The giant swallowtail caterpillar is one of a few species that elude birds by pretending to be their poop. What a fantastic skill, eh? A little humbling, perhaps, but worth it if it ensures your survival.

But don't think it's easy. It's not enough just to have the coloration of poop. No, you have to really play the part to be convincing. Some moth caterpillar species have white-and-brown coloration that gives them the appearance of a bird dropping. Yet they take things a step further by adopting the *posture* of poop. Talk about method acting! And how does a dollop of doo-doo look when resting on a leaf or branch? Well, the caterpillars rest in a curled position. It's very effective. And I'll be honest with you here: There's a long list of insects that are willing to look like poop to disguise themselves.

LET GO OF A LEG

Certain insects are willing to sacrifice a body part to avoid capture. Usually, it's a leg. You'll see this with some frequency in crickets, walking sticks, and leaf-footed bugs. A special muscle allows a leg or an antenna to snap off at the insect's will under the right circumstances. Built-in fracture lines at certain joints allow the appendage in question to break off cleanly, especially when in the grip of a predator. This phenomenon, known as autotomy, allows the insect to lose a piece of itself and save its life by distracting a hungry predator. When the predator stops to examine or eat the severed bit, the victim hobbles off. In some cases, the insect can regenerate the missing part later.

PLAY DEAD

Move along. Nothing to see here. Just a dead, dead, dead insect. I know it looked like it was alive when it spotted you, but then it fell off the tree branch and is now lying on the ground belly-up with its legs in the air. So dead. Right? Could be, or it could be something called thanatosis. It's surprisingly effective. It shuts a predator down; things that eat other things quickly lose interest in dead things.

The insect will remain frozen like this for some time, even when prodded or poked by the predator. After the predator has lost interest, the insect will start moving again and make its escape. The blue death feigning beetle (*Asbolus verrucosus*) goes belly-up when threatened.

Cuckoo wasps do the same thing. They curl up into hard, rigid balls. Click beetles (*Ampedus nigricollis*) drop to the ground, curl up their legs, and fold back their antennae. It looks like game over for these dudes. But they're all faking.[28]

Insects have a lot of defenses. Some insects employ more than one method. If you're an insect and you're ever in doubt over what to do, then I say this: Vomit. That's right. Just blow chunks. Because no one likes vomit, am I right? So if you're an upchucking insect, whatever it is that's chasing you or thinking you might be delicious for lunch is going to think twice. Grasshoppers know this. Grasshoppers are pukers. They will not hesitate to projectile vomit all over you. Whatever it's been chowing on all morning, it's coming back up in reverse. After that, what predator is going to say, "I want to eat that"?

You're probably exhausted. Me, I've got so many other wild things to tell you. Up to this point, I've been describing different species of insects and their great strengths, but not their power as a collective. There's a lot to say about swarms and colonies and armies. There's also a lot to say about how we depend on insects. Stay with me as we explore how insects feed the world in the next chapter.

---

[28] Not sure how you tell REAL dead from FAKE dead . . . Must be a duration thing?

2.

FEEDING THE WORLD

YOU'VE PROBABLY HEARD this before: Three-fourths of the world's flowering plants depend on pollinators to reproduce. And this little chestnut: One out of every three bites of food we eat exists because of animal pollinators. These oft-quoted estimates indicate just how vital pollinators are to planet Earth.

There's a lot to unpack when it comes to pollination—What is it? Why do we need it? Who is doing it? What happens if they don't?

Let's start with a little plant talk. You might think, *I'm not into flowers. Why should I worry about flowering plants?* Flowers are the things that help plants make seeds. Seeds carry all the genetic information to produce new plants. So we like flowers, and we like seeds because we need plants. We need plants to eat. We need plants to provide food and habitats for other animals. We need plants because they take in carbon dioxide and give off oxygen. Seeds good. Flowers good. Plants good.

For flowers to make seeds, which then make new plants, they need to be pollinated. Inside a flower, there are male bits—called stamen (it has a part called an anther, which holds the pollen)—and female bits—called pistils (the top part of the pistil is called the stigma, which collects pollen grains from pollinators). Pollination is the act of transferring pollen grains from the anther of the stamen of one flower to the stigma of the pistil of another. When pollen lands on the stigma, a tiny tendril extends down from the pollen grain and takes root in the plant's ovaries. Sperm rolls down this channel and fertilizes the plant's eggs. Most flowers have both anthers and a stigma and could self-fertilize, but the best seeds are formed when flowers are fertilized by pollen from different flowers of the same species.

It's a bit tricky for plants to make this magic happen on their own because most of them are rooted to the ground. They can't get up to spread their own stuff around; they need help. Scientists call this help vectors. Wind. Water. Animals. These are all good vectors to help pollinate plants. Wind can blow pollen to other plants. Rain can carry it too. Birds are great pollinators. We've got some crackerjack mammal pollinators like bats and lemurs, and we've even got some lizard pollinators. But the bulk of pollination is carried out by insects.

It's bees, right? Everyone knows that it's bees. Bees are the pollinators! Yes, yes, calm down. Bees are pollinators, but many other non-bee insects are pollinators too. We'll get to that in a minute. First, I do want to talk to you about bees.

Most people talk about bees like they are one thing, like there's one universal bee that does all the bee-ing in the world. The honey bee, right? Honey bees live together in hives, make honey, and sting people without mercy. The end.

Not exactly. If your knowledge of bees is hazy, that's okay. I'm not going to judge you. But I don't want you to sound ignorant when you talk about this with your friends and loved ones, so let's get you up to speed. First, there's something I should tell you, and maybe you should sit down for it. Good. Comfortable now? Ready? Here it is: There is more than one type of bee.

THE ABCs OF BEES

The bee you're probably most familiar with is *Apis mellifera*, the European honey bee—the go-to bee in the world of pollination and honey production. If I were to press you and you were more enlightened than the rest of us, you might be able to conjure up the names of other types of bees—bumble bees, sweat bees, carpenter bees, maybe? That's a pretty good list, but Earth is home to more than 20,000 species of bees. So the honey bee, the bumble bee, the sweat bee, and the carpenter bee, and . . . 19,996 *other* known species. North America alone supports 4,000 types of bees.

Bees can either be social or solitary. Honey bees are social bees. They form colonies and live and work together. But more than 90% of bee species are solitary. They don't live in hives, they don't hang out together, and they don't produce honey. Yet these bees are often excellent pollinators. If you already know this, you can skip ahead in the chapter to the part where I tell you how much pollinating is actually happening by non-bees. If you have never heard of the black-winged cuckoo orchid bee or the polysocial hairy-tongue bee or the vectored Sputnik bee, then hang here with me for a minute.

I'm always loath to say something is the biggest or the smallest. Invariably, as soon as you make a claim like that, someone somewhere will come along and tell you you are ignorant for not knowing that the blah-blah-blah species is *really* the biggest or the smallest. So I'm going to phrase this very carefully and say *to my knowledge*, the largest bee is the handily named Wallace's giant bee (*Megachile pluto*) and the smallest bee is *Perdita minima*.[1]

The mother of all bee species is named after Alfred Russel Wallace, an English naturalist who discovered it during an 1859 expedition. He was mucking around a group of Indonesian islands called the North Moluccas at the time. How giant is the giant bee? The females are about four times the size of a European honey bee. Their bodies are about as long as your thumb, and their wings stretch more than 6.3 centimeters (2.5 inches) across. Wallace wasn't really impressed with their size, actually. According to his journal, he was way more wowed by the bee's ginormous jaws, which he likened to a stag beetle's.

No one really knows much about this bee, other than it bunks down in termite nests and uses its mega-chompers to scrape up tree resin. After Wallace's discovery, the giant bee wasn't seen or heard of again until 1981—long after everyone thought it had gone extinct. Four guys went looking for it again in 2019 and captured sight of a live one in the wild. While that was big news, we still don't know much about it.

On the other end of the spectrum, you've got *Perdita minima*. We know quite a bit more about this bee. It's so small that ento-mologists have an easier time spotting its passing shadow than

---

[1] You don't have to take my word for it. You can go to a more trustworthy source: *The Guinness Book of World Records*. If you look it up, you'll get your confirmation. And while you're poking around the bee section, you'll probably come across a fella from Kerala, India, who holds the world's record for the "longest duration with head fully covered with bees." That's right: He spent 4 hours, 10 minutes, and 5 seconds with his head completely submerged in a cloud of bees. Why? Why? Why would anyone do such a thing?

spotting the actual bee. And forget about trying to net it; it can easily pass through the fabric of ordinary insect nets. I once saw a photo of this bee. It was perched on top of a carpenter bee. A carpenter bee is a tiny thing in and of itself, and *Perdita minima* was perched on its forehead.

These itty bitties are solitary bees, and they can be found in the southwestern deserts of the United States. While some bees are none too picky about where they get their pollen and nectar from, *Perdita minima* has a deep love only for the nectar and pollen of wildflowers in the Spurge family, such as the whitemargin sand-mat (*Chamaesyce albomarginata*). Which sounds especially delicious, doesn't it? It's a match made in heaven. The ol' sandmat's flowers are very small, and other insects have a hard time collecting its pollen. Though *Perdita minima* is tiny, its hairy little legs can carry quite a bit of pollen at a time.

Of the social bees, the most widely known are probably the honey bee, the bumble bee, and the Africanized honey bee. All are rather skilled pollinators. Bees end up in the role of pollinator by default because they frequent flowers so often. Their nutritional needs are almost entirely met by them. Bees eat a little bit of pollen and a lot of nectar. They feed their babies a lot of pollen and a little bit of nectar. You know how all this works, right? Nectar is that sugary liquid produced by flowers to lure pollinators. If you're interested in honey, then you know what nectar is. Honey bees glug, glug, glug the stuff. It goes straight to their honey stomach (yup, honey bees have two stomachs). When they get back to the hive, they barf up the nectar into the mouths of worker bees waiting at the hive entrance. Then the worker bees re-barf it up into honeycomb cells. This is the stuff you eat later, known as honey. Which is why you'll hear people say that honey is bee barf. I mean, yeah, *technically*.[2]

That's the story on nectar. Pollen is the breakfast of champions—an excellent source of protein, amino acids, carbohydrates, lipids and fatty acids, vitamins, and key enzymes. Bees are very motivated to go rooting around in flowers, and while they're doing so, they pick up a lot of pollen that accidentally gets transferred to other flowers.

ABOUT POLLEN

A lot of bees are fuzz butts. That's not really a scientific designation; it's more like BrennaSpeak. What *is* true is that a honey bee has about three million hairs on its body. How do we know this? Because some really smart folks at Georgia Institute of Technology figured it out for us during a study on bee hair and pollen collection.

I did ask, and no, they did not recruit people to sit around and count individual bee hairs. Dr. Guillermo J. Amador tells me that his team measured the width and length of a bee and scanned its body using an electron microscope. The scan measured the length, width, and density of the hairs. Then they took an average of the density for all the parts measured and used that value to estimate the total number of hairs. Knowing that some parts of a bee are more densely haired than others, they rescanned specific parts— like the legs and head—to get a more accurate count.

Which is how we know so much now about their hairy little eyeballs. *Eeew. Eeew. Eeew!* Their eyeballs are hairy? Yes, yes, they

---

[2] The process is called trophallaxis. It literally means the transfer of food by mouth from one individual to another.

are. And their leg hair is apparently five times more dense than their eyeball hair, so there's that. What Dr. Amador and his team discovered was that the hairs on a honey bee are *strategically placed* for maximum pollen storage. High-speed footage of tethered bees covered in pollen confirmed their theory. With all that hair, bees can carry 30% of their own weight in pollen. The only catch is that pollen has to be distributed in such a way that a bee can fly while it's carrying it, and the bee must also have clear vision to do the flying.

As it happens, Dr. Amador's study was the first quantified study of the honey bee grooming process. His team learned that bees have a pre-programmed cleaning routine, no matter how much or how little pollen they have on them. Bees always swipe their eyes a dozen times (six times per front leg) to clear their vision and move pollen down their bodies.

Dr. Amador's research found that the gap between each eye hair is about the same size as a grain of dandelion pollen. This distance between hairs is just enough to keep pollen grains suspended above the eye until the legs can comb through to collect them. When a honey bee lands on a flower, it uses its four front legs to scrape pollen, which it then transfers onto tufts of stiff hairs on its back legs. We call these pollen baskets. Why? I do not know. They aren't really *baskets*. More like stiff, dense patches of hair. They allow bees to store and transport pollen back to the hive. I used to think pollen was flyaway stuff like fairy dust, but a lot of it is actually pretty sticky. Bees sort of mash it together into balls that they stick to their baskets.

Here's something else that's cool: Bee hairs are electrostatically charged. The charge builds up as a bee flies. So some amount of pollen naturally clings to the bee even before it decides to slap any onto its thighs. As the bee moves from flower to flower, some pollen gets knocked off, helping to pollinate the next plant.

BUZZ POLLINATION

While honey bees do the lion's share of bee pollination, they aren't the perfect pollinators for all plants. Consider the tomato plant, if you will. Tomatoes are self-pollinating, meaning their flowers have both anthers and stigma, so only one plant is needed for reproduction. That's rather handy, isn't it? There's just one small catch.

A tomato plant's yellow flowers dangle down from the stem of the plant. As they bloom, the petals splay back, away from a central structure, which is a cone of five fused stamens. Inside the stamens, anthers produce pollen. The pollen, though, is sort of trapped inside the anthers and can only be released through tiny pores when you tap the plant. Think of shaking salt from a shaker. Like that. Once the pollen is released, it falls downward onto the pistil and voilà! Pollination.

Yet a honey bee simply does not have the heft needed to shake a tomato flower hard enough to release its pollen. No, what you need for that kind of work is a buster. What you need is a bumble because bumbles have a certain heft to them. They're also very fuzzy. Could there *be* a bee more follicly endowed? Possibly not. I'm not sure if anyone has ever counted a bumble bee's hair (probably?), but it's thick, plentiful, and branched at the ends too, which enables more pollen to stick to each hair. This hair is important, to be sure, but the magic lies in the bumble's brute force.

As it turns out, a bumble bee is a remarkably muscular creature.

Just ask Dr. Callin Switzer. He was working on his PhD at Harvard back in 2017 when he fell under the spell of the bumble. Officially, he was studying a phenomenon known as buzz pollination,[3] which is essentially what you call it when an insect like the bumble bee shakes a flower, such as the tomato flower, in order to release its pollen. It's something honey bees can't do.

Dr. Switzer was recording bumble bees to determine the frequency at which they flapped their wings and otherwise shook their bodies. It seems that bumble bees have some pretty impressive abs. Ha ha. No, I'm kidding. We don't call those abs. They are mesosoma—that middle section of an insect, also known as the thorax. Because of their mighty mesosoma, they can flap their wings about 190 times a second—that's about 50 times faster than it takes a person to blink just once.

Now, when a bumble bee gets down to business, it places its upper body close to a tomato flower's anther and bites down with its powerful jaws. Then it buzzes. Not just any buzz, though. It tucks its wings into the resting position and contracts its flight muscles. This causes the bumble's thorax to pulse in a rapid burst, which gives rise to a vibration that lasts about a second. This literally shakes the pollen from the anther.

Based on Dr. Switzer's recordings, he calculated an average buzz pollination frequency to be about 270 hertz. For those of us who have no idea what a hertz is, much less if 270 hertz is an impressive number, Dr. Switzer tells us that the bumble's buzz is equivalent to a C-sharp above middle C on the piano.

Buzz pollination coats the bee in pollen, which makes it happy, and allows pollen to drop down onto the stigma to pollinate the plant. For certain plants like tomatoes, potatoes, blueberries,

---

[3] The scientific word for buzz pollination is *sonication*.

cranberries, and eggplants, this may be one of the most efficient ways of releasing pollen.[4] The bumble is one of the few social bees that has enough heft to get the job done.

KILLER BEES ON THE LOOSE

That's not to suggest that other social bees don't help with pollination. Africanized honey bees certainly do. And while they are heroes to pollination, Africanized honey bees were saddled early on with an unfortunate and unfair nickname: killer bees.[5] When I first heard about killer bees, I imagined that they must be huge—like the size of your fist. They probably had fangs, and they drooled. And maybe even smelled bad or shot laser beams from their eyes. You know, something really terrible.

In truth, it's very hard to tell the physical difference between these bees and your standard, run-of-the-mill honey bee. They're nearly identical in shape, size, and coloring. They do not have fangs. Their sting is no more dangerous than other bees'. It's their behavior that is unusual: They are extremely territorial and much more aggressive than other bees.

These so-called killer bees were the brainchild of Brazilian humanitarian, geneticist, and scientist Warwick Kerr. With the best of intentions, Dr. Kerr crossbred the African bee with an Italian variety of the European honey bee in the 1950s.

For years, Dr. Kerr had watched Brazilian farmers struggle with

---

[4] Dr. Switzer also has done research on Australian blue-banded bees. It was assumed that they, too, would *shake, shake, shake . . . shake their booty* to release pollen, but no. High-speed videos revealed that these bees dislodge pollen by head-bashing flowers. The rapid-fire headbutts—350 times a second—create an audible buzz that knocks the pollen free. Who knew?
[5] Not to be confused with murder hornets (*Vespa mandarinia*), which are a lethal threat to honey bees

crop pollination. The European honey bees imported from Portugal at the time didn't do so well in Brazil's heat and humidity; they succumbed more readily to disease. African bees weren't great at producing honey, but they were more hearty. They bred and built their colonies up quickly. Dr. Kerr's notion was to build a better bee—a more successful pollinator—by crossbreeding the two.

Africanized honey bees would eventually become a key to his country's improved agriculture and honey production. As a nation, Brazil's honey production went from 43rd in the world to the 7th-largest honey producer. Today, most of the world's organic honey is produced by Africanized honey bees in Brazil's remote forests.

This new bee had some traits that made it an exceptional pollinator—a refined sense of smell, quicker movements, willingness to fly during bad weather, and superior navigation skills. But the bee Dr. Kerr created was far from perfect.

Africanized honey bees respond to any and all perceived threats with a large-scale response. An attack by these bees is vicious and sustained. They are capable of chasing people for more than a quarter of a mile to defend their hive. Victims don't sustain one or two stings; they suffer from hundreds of stings. That much bee venom all at once can send even a healthy, nonallergic adult into shock. Death is not an unlikely outcome.

Dr. Kerr was still tinkering with the breeding of these bees when disaster struck. One of his technicians mistakenly released the queens from 26 hives. The queens went on to form feral colonies, which have spread with unprecedented speed throughout Central and South America, Mexico, and now into southern parts of the United States. So much for a better bee. We might be grateful for their enhanced pollination abilities, but Africanized honey bees might be too much bee for any of us to handle.

SWEAT BEES DON'T SWEAT

Not all social bees are terrifying, but some small ones can still make us sweat. Halictid bees fit that bill. They are also known as sweat bees. These little dudes live in colonies underground or in hollow trees. I can remember getting stung by one when I was a little kid and thinking that it was roughly the end of the world. Getting stung by a sweat bee is not uncommon. They're fairly nonaggressive, but they are attracted to human perspiration, which is laden with salt. The bees seek that salt to meet their nutritional needs, so they like to land on our skin and lick the sweat off us. It's easy for them to get tangled up in our clothes, causing us to freak out, and they resort to stinging us. They die. We end up with an angry welt. Nobody wins; nobody is happy.

In researching the sweat bee, I discovered that they pollinate a lot of wildflowers and some crops like alfalfa, pears, and loquats. What the heck is a loquat? Do you know? I had to look it up because I'm pretty sure I'd never heard of such a thing. It's a fruit that comes from a large shrub. The fruit is sometimes called a Japanese plum. It's also called a Chinese plum, which must irritate both countries, having to vie for ownership like that. It originated in China but quickly took root, so to speak, in Japan and has grown there for about 1,000 years. Japan is now the leading producer of loquats with an annual crop of about 17,000 tons. Which frankly, is an awful lot!

The plant produces small, white, sweet-smelling flowers in the early fall. These flowers are then pollinated by sweat bees and others of that ilk. The fruits have a blended flavor of peach, citrus, and mango. I've never eaten one, but it sounds rather lovely.

The reason I bring this up at all is because, statistically speaking, a lot of people apparently eat the loquat. I don't know if they can't live without it or if it's a huge part of anyone's diet, but it's clearly a thing. I had no idea. And now that I do, well, it forces me to reassess the sweat bee, doesn't it? I feel bad for being all judgy. I had *one bad encounter* with *one sweat bee*. But maybe *I* was the problem? Maybe the bee had its head down, doing its important thing, and I got in its way? I could be thinking too much about it. It's a small thing, but I think small things matter.

THE SINGLE LIFE

Social bees aren't doing all the work, though. Remember, most bees are solitary, and solitary bees make a significant contribution to pollination as well. These kinds of bees live alone. They have a very small footprint and no social media presence to speak of. They don't enjoy large gatherings. They keep to themselves, preferring instead to stay at home and play endless rounds of solitaire.[6] They produce just enough food to feed their larvae, but they are dynamite pollinators. Of the solitary bees, you might be familiar with carpenter bees, mason bees, and leafcutter bees.

I don't want you to think that there's one type of carpenter bee. The genus includes some 500 bees.[7] So we can talk about them here, but just know that there's a lot of variation.

---

[6] Okay. *That part* I made up. I have no idea which card games they play.

[7] Always remember: **K**oalas **p**refer **c**hocolate **o**r **f**ruit, **g**enerally **s**peaking. That's a pneumonic for how you can remember scientific classification. **K**ingdom, **p**hylum, **c**lass, **o**rder, **f**amily, **g**enus, and **s**pecies. Carpenter bees are in kingdom Animalia, phylum Arthropoda, class Insecta, order Hymenopterans, family Apidae, genus Xylocopa, and species . . . Well, like I said, there are more than 500.

Carpenter bees are the buzz saws of the bee world. The females use their sharp mandibles to bore perfectly round holes into wood

to build their nests. Forget any "measure twice and cut once" precautions. A female carpenter bee knows exactly what she is doing. She carves out a hole about 1.2 centimeters (0.5 inch) in diameter, cutting against the grain. When the tunnel is about 2.5 centimeters (1 inch) deep, she banks left or right and keeps chewing her way along, this time *with* the grain of the wood. Don't believe any of those tales about her eating the wood. She isn't eating it. She's just excavating it. You'll see the telltale piles of wood shavings beneath the entrance to her nest.

Alas, the carpenter bee is not mindful of her shavings. She's more focused on digging out an intricate tunnel system with multiple chambers, where she can lay her eggs one at a time. Moving from the deep to the shallow, she leaves each egg with a supply of pollen before sealing it off with regurgitated wood pulp.

I digress here a little.[8] We're supposed to be talking about

---

[8] Talk about digressions . . . You probably didn't know this, but the Galápagos carpenter bee (*Xylocopa darwini*) is the only species of bee found in the Galápagos Islands. It is, by default, the archipelago's most important pollinator. It's endemic to the islands, and you want to know how they think it got there? (I love this story.) There are two prevailing theories. The first is that it flew there from South America—a journey of 1,000 kilometers (621 miles) or more. The second theory is that it *floated* there. In a pile of driftwood. I kid you not.

pollination. You really shouldn't let me wander off like that. Due to their short mouthparts, carpenter bees are important pollinators for open-faced or shallow flowers; for some, such as passionflowers, carpenter bees are the sole pollinating species.

Remember me telling you about bumble bees and buzz pollination? Carpenter bees have that ability too. For that reason, they make good candidates for greenhouse pollination. If you're a fan of the honeydew melon, you are a fan of the carpenter bee. The same can be said for blueberries and greenhouse tomatoes.

I recently read that insect pollinators such as honey bees contribute a value of around $29 billion to our agricultural industry—with about 15% of this value coming from native bees like carpenter bees. That's a pretty substantial contribution.

MAKING MUD

Mason bees (genus Osmia) don't look much like bees. For starters, they aren't black and yellow. They're metallic black and blue. They're smaller than honey bees, and their shape is more fly-like[9] than beelike. They don't act like honey bees, either. Mason bees are solitary, remember, so there is no queen bee. Rather, they are their own queen, as the beekeepers like to say. The males[10] live a shockingly short life. They spend about two weeks mating then die, leaving the females to a six-week lifetime of toil. The female's primary goal is to lay eggs and tend to them before she dies—a brief but focused existence.

---

[9] Listen and you'll be able to tell if it's a fly or a mason bee. Flies make a humming sound. Mason bees buzz.

[10] It's easy to tell the males from the females because the males have white noses.

Unlike carpenter bees, mason bees don't dig their own nests. They usually look for an existing spot that they can improve upon. This might be woodpecker holes, insect holes (like an abandoned carpenter bee's nest), or hollow stems. Something long and tubular is generally the criteria.

The female will forage for pollen and nectar and stash these things in the back of the nesting cavity until she deems there's enough to feed a young bee. Then she lays a single egg. Next she gathers up mud or clay and builds a wall[11] to seal off the egg. This is how she got her name—because of her masonry work. She repeats this process about five or six times in a single cavity before moving on to the next spot. In her lifetime, she may lay about 34 eggs.

The work of gathering food for her babies is pretty laborious. It starts with pollen. Honey bees, as we know, are quite fastidious about their pollen collection. They meticulously comb pollen off their faces and bodies and pack it neatly into sticky balls carried on their back legs (the aforementioned pollen baskets). Mason bees pollinate flowers with wild abandon. They don't have pollen baskets, but they do have hairy tummies. When they reach a flower, it's less of a controlled landing and more of a haphazard belly flop. As she's collecting food, a female mason bee crawls all over blossoms, dusting her tum quite thoroughly with pollen. When she belly flops onto the next flower, she ends up dropping a lot of that pollen, thereby pollinating the new flower.

Like honey bees, mason bees are hard workers, though. Each time she goes out on a foraging run, she's liable to visit 75 flowers.

---

[11] It takes an additional 10 trips or so out and back to gather materials for each wall.

She might make up to 35 trips to collect enough nectar and pollen to feed one larva. Do the math: It's a lot of trips. She visits thousands of blossoms every day. Mason bees don't go far to forage. Their range is limited to a radius of about 100 yards (think the length of a football field). What they lack in distance, they make up for in earnestness. Mason bees are willing to work on cool days or in rainy weather when honey bees are more likely to take the day off.

Masons visit fruit trees such as apple, plum, pear, almond, and peach. They are incredibly effective as pollinators—even more so than honey bees. The honey bee pollination rate is surprisingly low (5%), but a mason's pollination rate is about 95%. Just two or three females can pollinate a mature apple tree on their own.

CUTTING LEAVES

If you think mason bees are cool, you'll probably be similarly impressed by leafcutter bees (*Megachile rotundata*). Like masons, leafcutters are solitary, cavity-nesting bees. People who are in the know about leafcutters will say things like, "Have you ever noticed little segments cut away from the leaves of roses, lilacs, or other shrubs?" To which my answer is usually, "No." I wish I were that observant! Had I noticed such a thing, I doubt I would have connected it to leafcutter bees. These bees are named after their habit of . . . wait for it . . . cutting out circular pieces of leaves from plants. After nibbling the little bits off, they carry them to their nests. If you've ever seen a photo of this before, you'll think it's been Photoshopped because it seems so unlikely. Once at home, momma leafcutter will build a leaf-lined cocoon stuffed with pollen, nectar, and a single egg. She will fill each cavity with as many cocoons as she can.

Like mason bees, leafcutters drag pollen around too, largely unaware of the service they are providing to the world. Fortunately for us, they are generalists. They pollinate all sorts of things—blueberries, carrots, cranberries, melons, mint, onions, peas, and others, including something called vetch. Crown vetch. Hairy vetch. Sweet vetch. Vetch is not the most attractive name for a plant, but apparently they are from the legume/pea family, so now we know that. Above all, the leafcutter bee is a friend to the alfalfa plant.

I keep telling you how ingenious some of these plants are with their strategies to get pollinated, and alfalfa is no exception. It's rigged with a trigger. The lower petals of each of its flowers make an enclosure shaped like the keel of a boat, which holds the anthers and stigmas. When an insect lands on the flower, it "trips" the flower's trigger, transferring pollen to the stigma and dusting the pollinator in the process.

Honey bees, it seems, are too smart for the alfalfa. Like little insect ninjas, they learn to rob the flowers—collecting nectar without tripping the trigger. Leafcutter bees either aren't that bright or somehow enjoy being smacked around by the alfalfa plant. They get hit all the time but keep coming back for more. Which is great news for the alfalfa. Leafcutters trip 80% of the flowers they visit compared to honey bees, which only trip about 10%.

Because of this "never say die" attitude, leafcutters were purposefully introduced by the US federal government in the 1940s to increase pollination efforts of alfalfa and yield higher seed production.

Today, alfalfa is one of the most important insect-pollinated crops. Not because there's a big rush on sprout-eating but because cattle eat alfalfa hay. No alfalfa means no hay. No hay means no beef or milk. Alfalfa hay is also fed to sheep, goats, and horses.

Entire tomes have been written about bees, of course. And I could easily natter on about them all day. I'm dying to tell you about the vulture bee[12] (*Trigona hypogea*), but we really must move on. If you are super interested in bees—of course, how could you NOT be?—you should really get a copy of the book *Bees: An Up-Close Look at Pollinators Around the World* by Sam Droege and Laurence Packer. It's transformative. You will never think of bees in the same way again.

BEES IN TROUBLE

Clearly we can now see that bees, including honey bees, are essential to pollination. But here's where things get dicey. If you depend on a single species and something happens to that species . . . well, you're in trouble. And something *did* happen to the honey bee, didn't it?

Back in the winter of 2006–2007, beekeepers began reporting unusually high losses to their hives. Not all hives survive winter, it's true, but some keepers were reporting losses as high as 90%. Here's the unnerving part. Beekeepers weren't stepping out to find piles of dead bees at their door. No. Beekeepers were opening hives and finding no bees—well, the hives still had their queens and some larvae. They had plenty of food stores. They were just missing their entire workforce.

---

[12] These guys aren't interested in plants. They feed on rotting meat then make a sort of honey out of it to feed their young.

Worker   Drone   Queen

Each honey bee hive has three kinds of bees: the queen, the drones, and the worker bees. There is only one queen, and she is in charge. She sends out power pheromones that tell everyone else what to do. She stays in the hive, guarded by her attendants, and lays eggs. Lots of eggs. She's capable of laying up to 2,000 a day. The male bees, the drones, are . . . Okay. Let's just say the drones are very *focused*. Their entire goal in life is to eat and mate. That's it. They don't do anything else. The rest of the hive—upward of 60,000 worker bees—is made up of females, and they do the rest of the work. They tend to the babies, build the hive, clean the hive, forage for food, process the food, and haul out the dead.[13]

---

[13] At the start of bee winter in the fall, the worker bees have their revenge, though. They drive all the male bees out of the hive and block the entrance. If the males try to come back in, the females rip their wings off, tear their legs off, or decapitate them. It's the worker bees' way of saying: "Look, we have no intention of feeding you all winter long if you refuse to do any work."

If a hive loses its queen and can't replace her, it will eventually die because there is no one to lay eggs or tell the workers what to do. But if a queen loses her workforce . . . Well, nobody had ever heard of that. Ultimately, the hive fails because there's no one around to get things done.

Hives were losing their worker bees in great numbers. Eventually, the phenomenon was given a name: Colony Collapse Disorder.[14] What was the cause of this awful thing? There seems to have been a number of factors that, in combination, brought on collapse. We can think of them as the four Ps: pests, pathogens, poor nutrition, and pesticides.

FORMIDABLE FOES

Honey bees have always had to defend against parasites and pests. But in 1987, a new plague found its way into the US from Asia: *Varroa destructor*, the sworn enemy of bees. Varroa mites, to put it bluntly, are vampires—parasitic mites that feed on the blood of bees.

When the queen lays a worker bee egg in the cell of the hive, that cell becomes a brood cell. After 3 days, the egg hatches into a larva, which "nurse" worker bees tend to and feed. On the 9th day, the nurse bees cap the cell with wax. For the next 11 days, the bee larva transforms. On day 21, she chews her way out of her cell as a newly formed adult bee. It is during the period when a bee larva is in its cell but the cell has not yet been capped that varroa mites strike.

A female varroa mite sneaks into the hive and hides under a bee larva in an open cell. Once the cell is capped, the mite lays two to five eggs. Days later, the varroa eggs hatch and grow. The young

---

[14] This was a new name for a seemingly new catastrophe, but if you go back more than a century, you'll see occasional bee disappearances and dwindling colonies in other years (notably in the 1880s, 1920s, and 1960s). Were these cases of Colony Collapse Disorder too? It's unclear.

mites and momma mite emerge from the brood cell at the same time as the bee.[15]

For sustenance, momma mites latch on to adult bees in the way ticks latch on to dogs. In addition to trying to suck the bees dry, the mites can also inadvertently transmit an array of hideous viruses to their hosts, which we'll address in a minute.

If a hive is strong, the bees can beat back varroa. But varroa is no joke. Beekeepers have identified it as their most serious problem causing colony losses today. Sadly, varroa are not the only pests. Sometimes the bees must do battle on more than one front.

Small hive beetles (*Aethina tumida*) native to sub-Saharan Africa found their way to Florida in the late 1990s. By 2014, they had spread to 30 states. They are considered a secondary threat to hives, only causing serious damage if a hive is already weakened. They can put significant stress on a hive because the beetles can lay a lot of eggs and the eggs hatch quickly. Once inside a hive, the larvae can feed on pollen, honey, bee broods, and eggs.

Small hive beetles are difficult adversaries. Honey bees have a hard time killing them because their shells are too thick to sting through. Instead the bees resort to chasing them down into cracks and crevices in the hive. Once cornered, the bees stand guard to imprison them. That should be the end of the matter. Unfortunately, the beetles tap the mouthparts of worker bees with their antennae, similar to drones begging for food, and are able to trick their guards into feeding them. This behavior allows the beetles to survive in confinement for extended periods.

A third pest threat comes in the form of wax moths. There are

---

[15] Drone (male) broods are capped for longer than worker (female) broods, so on average, more varroa mites are able to mature in drone brood cells.

two species that wreak havoc on hives: the greater wax moth (*Galleria mellonella*) and the lesser wax moth (*Achroia grisella*). Both species eat beeswax and pollen. They can appear in active hives, but they have a field day with unattended combs in storage. They are the bane of any beekeeper who tries to store hive frames in a dark, warm, and poorly ventilated place like a garage or storage shed. Which, frankly, is where most beekeepers keep things like that. Rarely can wax moths topple an active hive on their own, but they can cause real damage to bee colonies that are already weak.

A beekeeper might be so busy trying to stamp out varroa, hive beetles, and wax moths, they might not initially pick up on signs of deadly pathogens in the hive. There are some particularly nasty ones out there, I'm afraid: deformed wing virus (as bad as it sounds), Israeli acute paralysis virus (which manifests as darkened and hairless abdomens and thoraxes, paralysis, then death), European foulbrood[16] bacteria (evident when dead and dying larvae appear discolored and curled up), and *Nosema ceranae* fungi (think chronic bee diarrhea). Treating these pathogens can be challenging at best if they are caught early, or they can decimate a hive.

Poor nutrition also poses a threat to honey bees. Just like us, if a bee's diet is poor, it's not going to feel especially well, and it's not going to have a strong immune system. Bees need pollen and nectar variety, but as more and more land is lost to development, we are losing habitats with diverse foraging plants.

Ugh. This is so depressing. We still have to cover the fourth P: pesticides. I'm sorry. But you need to know. Pesticides are poison. Poison is not good for living things, obviously. There's an entire

---

[16] There is no coming back from this. When it is discovered in a hive, the entire hive must be burned and all the bees destroyed. It is a terrible, terrible thing.

class of pesticides called neonicotinoids that can be quite damaging to bees. Their names look like a string of typos—clothianidin, thiamethoxam, dinotefuran, and imidacloprid.

These nicotine-based pesticides were developed in the mid-1990s because they didn't seem to harm wildlife as much as other pesticides. Hardly a ringing endorsement, I know. What we're finding, though, is that honey bees exposed to even nonlethal levels of this stuff can experience serious problems with flight and navigation, reduced taste sensitivity, and cognitive deficiencies that impede their ability to forage. In short, it's bad for bees.

Honey bees might be able to survive many of the four Ps if these problems were to occur one at a time. But when they hit in combination, they can destabilize a colony and cause its collapse. While I've presented it to you in a fairly concise passage here, it has taken years of study and research from some of the top minds to figure this out.

WHO IS POLLINATING?

While all the apiologists[17] of the world were frantically trying to sort out what was happening with the honey bees, there were a number of scientists asking a different set of questions. They challenged the notion that pollination was all up to the bee. They asked, *Which other insects pollinate? How much do they pollinate? And if all is lost for the poor honey bee, can something else take over so we don't all starve to death?*

Here's where science can get really interesting because we don't know what we don't know, right? If something hasn't been studied, then how do we know if we can count on it or not?

---

[17] A person who studies honey bees is an apiologist. A person who keeps honey bees is an apiarist. A person who studies all types of bees is a melittologist.

Dr. Romina Rader of the University of New England wanted some answers on this bee/non-bee pollination issue. She put together a multinational team of scientists to look at pollination studies of non-bee insects. What was up with the flies, beetles, moths, butterflies, and wasps of the world? Dr. Rader and her colleagues analyzed data from 480 fields for 17 crops in 39 studies on five continents. And you know what they discovered? Non-bee pollinators performed around 39% of the total flower visits. In most cases, non-bees weren't as effective as bees during a single visit, but they made more visits and they visited several plants that bees typically ignore. Dr. Rader's team also found that non-bee pollinators were less sensitive to habitat fragmentation than bees were. This gave researchers a bit of hope: Non-bees might be a bit of an insurance policy against honey bee declines.

FROM THE PLANT'S PERSPECTIVE

Before we talk about what else is out there in terms of pollinators, I want to double back to plants again. Here's something that bears remembering: The key to pollination is the plant not the pollinator. What do I mean by that? From the pollinator's perspective, pollinating is accidental or, at best, coincidental. It's not the insect's primary goal. It's visiting flowers because it wants food in the form of nectar or pollen. But from the plant's perspective . . . I mean, if plants *have* perspectives . . . pollination is everything. It's survival.

As a result, many plants have evolved in certain ways to attract pollinators—in some cases to attract a very specific pollinator. The folks who study this sort of thing even have a name for it. They call it pollination syndrome. It means that plants and pollinators have coevolved physical characteristics that make them more likely to interact successfully. It means that it's a win-win situation for both plant and pollinator.

In the case of a honey bee, flowers with the most success will have the following characteristics: The petals will be bright white, yellow, blue, or ultraviolet (honey bees can't see red; if the flower has red petals, a honey bee will likely skip it). The flower's shape will likely be shallow with a landing platform for the bee to rest on. Its pollen will be sticky. It will have a fresh scent. Luckily, lots of flowers fit the bill for the honey bee—lavender, buttercup, aster, clover, sunflower . . .

But as we've pointed out, honey bees are not the only pollinators. Which other pollinators come to mind? If you ask the average person on the street to name an insect pollinator, the humble bee would probably be top of mind. The next insect on the list would likely be a butterfly.

TO LURE A BUTTERFLY

They're so visible, those butterflies. They're active during the day. Many have beautiful colors. They visit colorful flowers. They are gentle to interact with. Butterflies take us to our happy place. Happy, happy, happy. There are no evil butterflies. They do not bite. They are not pests. Who has ever had a harsh word for a butterfly? You'll never hear a person say, "It would have been a perfect picnic were it not for those noisy butterflies!" No one ever says, "The wedding was ruined by the release of those graceful butterflies!" No. By and large, people have good associations with butterflies. Especially monarch butterflies. People easily fall into delirium when they talk about monarchs.

And why not? Monarchs reign. They're attractive. They have a great backstory. They go on long, epic journeys. Scientists have a ton of data on their lifestyles, their travel plans, their vacation spots, their dining habits. Milkweed. Everyone knows about milkweed. We all know to plant it, to fill our yards with it because monarch caterpillars exclusively feed on the leaves of milkweed plants. The milky juice of the plant makes the caterpillars and the adult butterflies distasteful to birds and other predators. It's fuel and protection rolled into one.

The monarch may be the most familiar butterfly, but the world is home to about 17,500 species of butterflies. When they are looking for flowers, butterflies tend to go toward plants that flower in clusters and open during the day, when they are also most active. Butterflies have a weak sense of smell, so the flowers don't need to be excessively pungent, but they do have to look good. A butterfly's vision is exceptional. They favor brightly colored flowers and, unlike honey bees, are able to see red.

To attract butterflies, plants often advertise themselves using nectar guides. Nectar guides are patterns that direct pollinators to nectar and pollen—sort of like that person on a runway waving around the orange, cone-shaped flashlights, leading the plane to a safe landing. Nectar guides often appear in ultraviolet (UV) light, which our eyes can't detect. UV light is picked up through special photoreceptors, which are the light-detecting cells in color vision. Like people, most insects have three classes of photoreceptors. Butterflies generally have four. The common bluebottle butterfly (*Graphium sarpedon*) has a whopping 15.

When we look at a black-eyed Susan (*Rudbeckia hirta*), we see a solid yellow flower. A butterfly sees a dark circle surrounded by a bright ring on the outer parts of the petals—a strong visual clue to head to the center of the flower for nectar. Sunflowers, daisies, and dandelions work the same way. Butterflies respond to the color of the petals not just for *where* to find nectar and pollen but sometimes for *when*. The color of the nectar guide of the horse chestnut tree (*Aesculus hippocastanum*) changes from yellow to red when nectar is no longer in production.

I never fully understood why people link butterflies so heavily to pollination. Don't get me wrong. Butterflies have some awesome superpowers when it comes to interacting with flowers, but they aren't great pollinators. They have a few structural drawbacks.

You've probably never given any serious consideration to a butterfly's legs, have you? Well, no, of course not. That's probably never been on your mind. But if you pause a moment with me now to consider their legs, you might realize how long they typically are. While they visit many flowers—rustling about in pollen and lapping up nectar with their long proboscises—their tall legs keep their fuzzy bodies hoisted fairly high above flowers. Butterflies can rob flowers of nectar from a distance and end up making less contact with pollen than, say, a bee does. Butterflies also lack any special structures like pollen baskets to cart pollen off with them. So while they inadvertently collect *some* pollen and transfer *some* pollen to other plants to pollinate them, less of that is going on than what you might suspect.

MOTH MAGIC

If you want the truth, it's really their insect cousins the moths that do more pollinating. Moth species outnumber butterflies 10 to 1. And they, by their very nature, tend to go after plants that butterflies aren't interested in. Before I tell you about their pollinating powers, I feel compelled to disabuse you of a few faulty notions you might be harboring against the moth.

As they are creatures of the night, most of us have little interaction with them. We don't really know them. We tend to associate them with holes in our woolen sweaters or con-nect them to the gaggy smell of mothballs—designed to repel them from our precious woollies. We malign them. If we refer to something as moth-eaten, it's not good. But I'm here to tell you that moths, speaking in the technical language of scientists, are "pretty freaking amazing creatures." And superfine pollinators to boot.

This idea that moths are drab and boring. Where did it come from? I can get pretty upset about this, but fortunately, moths are easily defended. For sheer size and magnificence, you must look to the giant Atlas moth (*Attacus atlas*). Its wingspan stretches more than 25 centimeters (10 inches). The white witch moth (*Thysania agrippina*) has the Atlas beat for wingspan, though—a full 36 cen-timeters (14 inches) across.

For color, oh, where do I begin? The elephant hawk moth (*Deilephila elpenor*)? Stunning! Pink and tan. So named because its wings spread out like two large elephant ears, and the pink spots on its body look like eyes. Another splotch of pink trails down the body like a long trunk.

Check out the coffee bee hawk moth (*Cephonodes hylas*). It's a moth, for sure, but it looks like a cross between a butterfly and a cicada. Its head is covered in lime-green fuzz. It has a band of deep red on its thorax, and its abdomen is splashed with bright yellow. Its wings are translucent so as not to appear too ostentatious—a subtle nod to fashionable restraint.

What about the rosy maple moth (*Dryocampa rubicunda*)? Have you ever seen such a thing?[18] A lush, canary-yellow body with

---

[18] Don't be fooled by this sweet-looking thing. Its caterpillar—the greenstriped mapleworm—is the foe to all maple trees. In dense populations, these caterpillars are capable of defoliating entire trees.

pink-and-white-striped wings. There are so many others—the garden tiger moth (*Arctia caja*), the lime hawk moth (*Mimas tiliae*), the Io moth (*Automeris io*). Have you seen the spectacular giant leopard moth (*Hypercompe scribonia*)? It's black and white, but its underbelly, head, antennae, and feet are tipped in cobalt blue.

For shock and awe, take a look at the death's-head hawk moth species (*Acherontia atropos*, *Acherontia styx*, and *Acherontia lachesis*). They each have a human skull–shaped pattern on their back! Explain that one to me.

If eerie isn't your bag, and you're looking for something adorbs instead, look no further than the Venezuelan poodle moth (*Artace* sp.). You'll think I'm making this up. Everyone thought this was made-up. In 2009, zoologist Arthur Anker nearly broke the internet after posting cute photos of this alleged moth that he spotted in Venezuela's Canaima National Park. But surely it was a hoax? Someone having fun with Photoshop? Next to nothing is known about this moth, but it does appear to be for real. It's probably a newly discovered species of silk moth.

If you're not into cute—and what's wrong with you, anyway?—I will leave you with one last example. This one, you'll just have to look up because I can't properly describe it: the arctiine moth (*Creatonotos gangis*). You tell me what's going on with the back end of this moth. Are those . . . tentacles? What *are* they? I mean, whatever they are, they are as long as the moth's whole body. Scientists think they are tubular scent glands[19] that inflate when the male of the species is hoping to mate. These scent organs produce a pheromone—hydroxydanaidal. I couldn't tell you what it smells like, but it attracts female moths.

NIGHT SHIFT

Okay. I think my point has been made: Moths—not so boring, am I right? Now! Back to the task at hand, which was trying to explain to you the role moths play in pollination. Most moths, you'll come to learn, are nocturnal or crepuscular. You probably know that nocturnal means they are active at night, but I had to look up *crepuscular*. It's kind of an icky-sounding word, I think, like some sort of wound that has scabbed over. It's nothing like that at all; it just means active during twilight.[20]

Remember, I was explaining earlier that the way a flower is built will help determine which pollinators flock to it? For moths, those flowers are typically night-blooming. That makes sense. The flowers bloom in clusters and some provide a nice landing platform for moths to pull up alongside and dock, as it were. Petals are white or dull colors—easy to see in moonlight. These flowers don't generally have nectar guides because color vision doesn't work as well at night. They produce a strong, sweet odor and a lot of nectar, which is deeply hidden so the moths have to burrow in to get their prize. Many moths are generalists. Some typical moth flowers are jimsonweed, stephanotis, honeysuckle, morning glory, gardenia, and tobacco.

---

[19] Three words that should never be used together: *Tubular. Scent. Glands*

[20] I guess "twilight" can be broken down even further. The term *matutinal* is used for animals that are active only before sunrise and *vespertine* for those active only after sunset. I do hope you will add these words to your daily vocabulary.

Among the more important moth pollinators are the hawk moths (family Sphingidae). These fast and aerobatic flyers have long, narrow wings and thick bodies. Some hawk moths are mistaken for hummingbirds because they hover in midair[21] while they feed on nectar. Their flight abilities have been closely studied, especially their sideslipping—moving rapidly side to side while hovering. This may have developed as a way to confuse ambush predators. Hawk moths need to suck down a lot of nectar to fuel their flight, and they know where to go to get that. Orchids.

A lot of orchids have unique relationships with their pollinators. The star orchid (*Angraecum sesquipedale*) is one such example. It was discovered in 1798 and described in 1822 by Louis-Marie Aubert du Petit-Thouars. He-of-the-Many-Names was a French botanist known for his work collecting and describing orchids from the islands of Madagascar, Mauritius, and Réunion. The lovely star orchid has a nectar tube 30 centimeters (12 inches) long. Our French botanist found many types of similar orchids that had yet to be named. English naturalist Charles Darwin was sent several samples to study. Of one sample, Darwin predicted there must be a specific insect capable of reaching its nectar and pollinating it at the same time. He famously wrote in the postscript of a letter to a friend: "Good Heavens what insect can suck it?" A question, no doubt, that we all would have asked.

Darwin's working theory, which he later described in a book on orchid pollination, was that an orchid such as this had coevolved with a specific pollinator. He believed that to be a type of moth with an especially long proboscis. Of course, he was mocked

---

[21] This hovering capability is only known to have evolved four times in nectar feeders: in hummingbirds, certain bats, hover flies, and these sphingids.

ruthlessly by the scientific community at the time for saying such a silly thing. But Alfred Russel Wallace—of giant bee fame mentioned earlier—was paying close attention. Wallace later published an article supporting Darwin's idea, going so far as to say, "That such a moth exists in Madagascar may be safely predicted; and naturalists who visit that island should search for it with as much confidence as astronomers searched for the planet Neptune, and they will be equally successful!" Hear, hear!

Of course, both men proved to be right. Years after Darwin's death, a moth was discovered in Madagascar with a proboscis that's 26 centimeters (10 inches) long. Unknown to science until then, it was named *Xanthopan morganii praedicta*. *Praedicta* is Latin for "predicted." Today it is commonly called Wallace's sphinx moth.

I merely point this out to you to further my case about the special relationships between pollinators and their plants and to raise moths ever higher in your esteem. How can you not feel that way now that you know moths were instrumental, nay, pivotal, to Darwin's and Wallace's great works?[22]

Hawk moths are known for their ability to travel incredible distances, which is great news for a flower. The farther its pollinator travels, the farther it spreads the pollen that's been dusted all over its fuzzy little face or body. Hawk moths tend to move pollen farther than bees or birds. Some species can travel as far as 29 kilometers (18 miles) on their feeding routes.

Hawk moths don't largely pollinate food crops—unless tobacco counts—making them less popular than insects that help the agricultural industry. But moths are vital for the survival of many native plants. For example, the red-flowering Puerto Rican higo

---

[22] Okay, okay. "Pivotal" might be laying it on a bit thick . . .

chumbo cactus. It lives on three small islands off the coast of Puerto Rico. To survive, it needs pollinators that can fly across the ocean. Hawk moths can make that distance and hop across island chains. Moths also pollinate the spiky Eggers' century plant, an imperiled species of agave that survives in small, scattered populations on Saint Croix of the US Virgin Islands. I know you've never heard of these plants and possibly don't care, but these plants depend on moths. They aren't the only plants to do so.

YUCCA DO, YUCCA DO

For millions of years, the yucca plant and the yucca moth[23] have been locked into a vital partnership. Neither can live without the other. The moth's larvae depend on the seeds of the yucca plant for food, and the yucca plant can only be pollinated by the yucca moth.

When I was very young, my family had a row of yuccas planted just beneath the window of the bedroom I shared with my older sister. On summer nights, that window would be open, and a cool breeze would pass through. When my dad tucked us in, he would tell us that the yucca plants had a song they would sing to the moths to bring them around each night. *Yucca do, yucca do, yucca do, yucca do.* If we were *very* quiet, he said, and stayed *very* still, we might be able to hear their song. I can remember lying there, waiting. Straining to hear that song. Hoping that if I heard it, I might be able to leap up and catch a glimpse of the moths arriving. But being so quiet and so still, sleep would usually overtake me instead.

---

[23] There are a number of species of yucca plants, each with its corresponding partner, a species of *Tegeticula* or *Parategeticula* moths. For example, in the central United States, soapweed yucca (*Yucca glauca*) is pollinated by the *Tegeticula yuccasella* moth.

Today, my dad disavows any memory of this, but my sister can confirm it. I've no idea how he came up with this song or even why. He probably just wanted us to go to sleep. As it is, the yucca plant *does* call to its moth. It just does it through scent, not song.

Each spring, adult moths emerge from underground cocoons, and the males and females meet up with each other on yucca plants to mate. When a female is ready to lay eggs, she first goes to a yucca flower to collect pollen. Yucca moths don't have proboscises like their moth brethren. Instead, they have two short tentacles near their mouth that they use to scrape pollen from the anthers of a flower. If you look them up and zoom in on their mouths, you'll see what I'm talking about. The female collects a sticky ball of pollen, which she tucks under her chin, then she flies off to another yucca flower.

When she arrives at the second yucca flower, usually one that has very recently opened, she goes straight to the bottom, opens a small hole, and lays her eggs inside. Then she scrapes a small amount of pollen from the tentacle ball, walks to the stigma of the flower, and packs the pollen into it.

Before she leaves the flower, she marks it with a pheromone to tell later visitors that they're not the first to reach the flower. Other moths will either lay fewer eggs than the first moth or none at all, depending upon how many moths have left their scent already. This helps moderate the number of larvae hatching within each flower and prevents the plant from aborting the flower altogether, which it will do if too many eggs are laid.

When the eggs hatch, the larvae feed on yucca seeds within the fruit. Typically, there are more seeds than the larvae in a particular flower can eat. When the larvae finish eating, they burrow out of the fruit and down into the ground to make their cocoons and wait until the next spring, when the whole process plays out again.

As far as anyone knows, and it's been studied since the 1870s, no other species besides the moth pollinates yucca flowers. Similarly, yucca moth larvae don't feed on anything other than yucca seeds. Each species depends upon the other for survival, and both benefit from the relationship. While this relationship is well-documented, many moth-to-plant relationships are lesser known.

MORE ABOUT MOTHS

But don't take my word for it. Dr. Richard E. Walton is in the know. He and his team from the University College London conducted a pivotal study on moth pollination in Norfolk, England, in 2016 and 2017.

I can report their findings to you. But before I tell you *what* they discovered, I want to first explain to you *how* they did their study because the *how* is really important here. This will become clear to you in a minute.

If you read their paper—"Nocturnal Pollinators Strongly Contribute to Pollen Transport of Wild Flowers in an Agricultural Landscape"—published by the journal *Biology Letters* in May 2020, the objective was to ask and answer the question: "How much do moths pollinate?" One way to get at an answer to that question was to measure the pollination efforts of bees, wasps, hover flies, and butterflies in a given area as they pollinate during the day and then compare those results against the pollination efforts by moths from the same area at night. Their research focused on the edges of nine ponds surrounded by crops and hedgerows in Norfolk farmland. The team observed and recorded daytime pollinators as they visited flowers. At night, they used light traps to lure moths into buckets set next to the ponds. In the morning, they took the moths to the lab to identify them and check them for pollen. That's the gist of it.

Before even reading their results, I had a lot of questions. Like, why nine ponds? Which nine ponds? How did they record the insects pollinating? How did they determine which pollen came from which flower? So I asked Dr. Walton a lot of things. Thankfully Dr. Walton answered every question put to him so that I could relay it all back to you. Are you ready? Here goes:

The ponds were chosen based on the types and amount of crops grown nearby and their location in relation to human foot traffic. Can't have a lot of pesticides floating about that might kill all the insects, now can you? You also can't have people traipsing through your study areas with their happy, pond-frolicking dogs. No. You need quiet, undisturbed spots. Dr. Walton knew these nine ponds fit the bill, as they had been part of a larger, earlier study on biodiversity.

Yet, before he could get too far down the road with all this, Dr. Walton needed to know what, specifically, was there to attract pollinators. That meant he had to create a pollen library of sorts. Each month, he walked the ponds and surveyed them. He'd take samples of each flowering plant. During the winter months, there wasn't much to collect, but during the spring and summer months, he was up to his eyeballs in pollen. He'd then take his samples back to the lab and transfer the tiny grains of pollen onto slides. Using a microscope and existing references, he began the painstaking work of correctly identifying and labeling each one. This took days. When finished, he added his slides to his growing collection in order to build a full pollen library of the nine ponds.

To keep track of his daytime pollinators, Dr. Walton used time-lapse photography. He set up two cameras at each pond to capture one image every 30 seconds for as many as 11 hours a day. So that. Times nine. This effort yielded thousands of photos per camera. Each image had to be reviewed to see if a daytime pollinator was present and, if so, to determine if it could be identified. And that's just to keep track of the day crew.

For the nocturnal pollinators—our friends the moths—a different protocol was required. These were caught with light traps and euthanized. I'm sorry to say that, but the team needed to examine the moths to see what exactly they had been pollinating and this was the only way it could be done.

You and I now know how fuzzy moths are. We know that Dr. Walton and his team meticulously combed through the moths' bodies looking for pollen. But there's another way scientists look for pollen: They check the moths' mouths.

I want you to picture a moth. Picture how small it is. Picture the moth's teeny-tiny, curled-up proboscis. Do you have that image in your mind? Good. Now picture yourself unfurling that thing so that you can take a pollen sample.

It's even more challenging than you think. To get a good sample, moths really need to be relaxed. *Ha ha*, you say. *How much more relaxed could they be? They're dead.* Yes, yes, I know that part, but their bodies harden after death, which makes unrolling that long tongue of theirs rather difficult. So what needs to happen? A moth gets pinned to a foam base that keeps its body upright. A mixture of water and vinegar is placed in the box to be absorbed by the foam. The box is covered to create a humid environment that relaxes the muscles and tendons of the moth's mouth.

Once the moth is ready, you dip the tip of a very thin, fine-pointed pin into a glycerin solution mixed with dye. With a steady hand, you gently, slowly uncoil the proboscis. Careful, careful, careful. I know you're probably thinking that you just eyeball some pollen and then smear it onto a slide, but, again, it's not that simple. The proboscis isn't a hollow tube like a straw. It's more like two straws hooked together. Pollen can potentially get stuck in between the two straws, so you need to get the pin tool between the straws to search for it. Once you find some, you have to collect the right amount. Too much, and your slide will be gloppy. Too little and you won't have enough for the slide.

Pollen from the body was easier to see and collect, but it had to be put onto separate slides so Dr. Walton and his team could identify how the moth was carrying pollen. This would help them later to understand the moth's overall strategy in collecting and transporting pollen.

Once a proper slide was made—and there were well over 1,000 slides per moth—the pollen had to be identified. The pollen library that Dr. Walton initially made helped with that. Identifying the pollen was one step. Each pollen grain also had to be *counted*. It was the only way they could tell how much pollen a moth was carrying. It took the team weeks to process a single moth.

Not to make your head hurt or anything, but the team had one other significant challenge during this process. I know, I know . . . as if trying to see these tiny creatures and their anatomy wasn't challenging enough. They were also sweltering. You see, it was very important to keep everything sterile. They couldn't risk any contamination or cross-pollination among the specimens. That meant the team had to work in a sealed space, trying not to overheat in their lab coats and gloves.

Those were the procedures. Think about all that. Now multiply it 838 times because that's how many moths the team swabbed and processed.

The report that the team generated was illuminating, including some fairly impressive charts that would make a normal person's eyes bleed. What did their work reveal? Of the 838 moths, the team found that 381 of them transported pollen from 47 different plant species, including at least 7 rarely visited by bees, wasps, hover flies, and butterflies. Meanwhile, a network of 632 bees, wasps, hover flies, and butterflies visited 45 plant species. And 1,548 social bees visited 46 plant species.

The moths tended to visit the same types of plants that the daytime pollinators did but were carrying heavier loads and making more frequent trips. A full 81% of the pollen-carrying moths carried more than 1 grain of pollen on their bodies, 30% carried 2–5 grains, 51% carried more than 5 grains, and 19% carried more than 10 grains. They also were more diverse in their selection of plants than the daytime pollinators; their food webs were fairly complex.

There's a couple of important things we can draw from the team's work. First, because the moths interacted with a lot of the same plants as the daytime pollinators, it suggests that moths could fill in the gaps should anything happen to our friends the bees. Second, we also now know what has long been suspected: Moths pollinate flowers in much the same way the daytime pollinators do. Fifty-seven percent of the pollen grains found on the moths were found on their fuzzy bodies. Finally the study helps us realize that we don't have a good baseline on moths. We don't have as strong a body of knowledge on moth pollination as we do for other pollinators. That's a gap that we should try to fill.

I asked Dr. Walton about this. Why no love for the moths? It's less that and more having to do with funding, he told me. Scientific research is expensive. Covering the costs of field equipment, field work, lab work, and training sessions can add up fairly quickly. Organizations that have the funding to back research projects want to make sure that there is a return on their investment. Studies directly related to understanding crop pollination, for example, are easy to get behind. Studies that ask questions about the little-understood relationship between flowers and night insects that people don't normally pay attention to can be a little tougher to sell. Or they *have been* tougher to sell in the past. Perhaps that's beginning to change.

TRY IT! IT'S FUN!

After grilling Dr. Walton for endless details about how he conducted his research, I began to wonder what it might be like to walk in his wellies. I didn't have any ponds nearby but do live directly across from a national park. I began to wonder what I might find if I went looking for moths. Before I could stop myself, I ordered a couple of black lights. When the lights arrived a few days later, my husband saw me brandishing them and merely raised his eyebrows. He knew I was up to something weird; he was probably just too exhausted by me to guess what it was *this time*.

My next act was to string a cord up in our backyard between two trees and fling a white bedsheet across it. Then I placed another white sheet on the ground in front of it. Dragging a long extension cord through the backyard, I connected my two black lights to shine onto the hanging sheet. I stood back to inspect my work. It looked like a nice-enough setup. The question was: Would anyone bother to show up?

I had picked a particularly steamy July evening for my first attempt. Around dusk, I flipped on the lights and waited. At first, nothing much happened. My family watched from inside, sadly shaking their heads. *There she goes again . . . doing weird things. This time in the yard. What will the neighbors say?*

As it grew darker, I could just begin to make out some small specks on the white sheet close to the ground. I crept up to the edges of the sheet and was stunned to find all sorts of bizarre things clinging to it. I'm probably the least-qualified person in the world to say what they all were, but I could identify some basic things like ladybugs and fireflies. In general, I saw moths. And beetles. A lot of beetles. I immediately started documenting as much as I could with my cell phone. To get decent pictures of small-scale things, you have to get fairly close.

Several hours later, my husband went looking for me. He didn't have to go far—he found me on my hands and knees with my butt in the air and my face inches away from the ground cloth.

"Are you coming in anytime soon?" he asked. "You've been out here for hours."

"Yeah, yeah," I said, "soon. But check out this cool thing with the doodly-doos on its head! And *this* one is bright green. And I think *this* might be some kind of parasitic wasp. And *this* one . . ." Knowing that this one-way conversation might go on for an extended period of time, Chuck backed away slowly and left me to my observations.

I was hooked. I was absolutely fascinated that all this stuff crawled up onto or flew into my sheets. If you ever want to try this, it isn't hard. Seriously. If I can manage it, *anyone* can—and you can do it just about anywhere, even if you live in a city or don't have access to a backyard. Hang up that white sheet and black light. If you build it, they will come, my friends. And I promise you: You will be amazed at what arrives.

COLLECTING DATA

What was I looking at exactly? Dr. Walton knew, of course. He rattled off the names of everything I showed him. Sadly, we don't all have access to a Dr. Walton or even *my* Dr. Walton. But we do have a pretty nifty tool that you may have already heard of. It's called iNaturalist.[24] This is a free app that you can download to your phone. When you are out and about and you see something amazing (or even quite ordinary) in the plant or animal kingdom, you take a picture of it and upload it to iNaturalist. You indicate the location of where you found said amazing/ordinary thing, and the app will give you suggestions on what it might be. Then you share your find with the community, and experts can confirm for you what it might be.

This thing is really easy and fun for regular people like you and me, but it serves a greater purpose. iNaturalist is a citizen science project built on the concept of mapping and sharing observations of biodiversity across the globe. The observations you and I record provide valuable open data to scientific research projects, conservation agencies, other organizations, and the public. Even if you know exactly what something is and don't need help identifying it, recording its presence in that location at that point in time helps

---

[24] inaturalist.org

build a powerful data set for scientists. It goes back to what Dr. Walton and others have been telling us. The more information we have on something, the easier it will be to understand its role and to assess its relative health. So far iNaturalist has logged almost 100 million observations.

Dr. Walton is right to collect as much data as he can about moth pollinators, and anything that you and I can do to add to the body of knowledge on this would be helpful to scientists. Because we're trying to piece it all together, you see. There's so much we don't know because we don't have data for it—or we think we know but we're merely guessing because we don't have evidence. Here's a good example of what I mean: the ghost orchid (*Dendrophylax lindenii*).

LOOKING FOR GHOSTS

Unless you frequent steamy swamps in Florida or Cuba, you'll be hard-pressed to see a ghost orchid in the wild. Even if you do frequent steamy swamps in Florida or Cuba, you'd have trouble finding one, partly because they thrive in remote, marshy places you wouldn't necessarily want to go to, partly because they are endangered, and partly because, unless they're blooming, they're devilishly hard to spot. They only bloom for a week or two, possibly during the months of June, July, or August. If they bloom at all. Our best guess is that as few as 10% bloom in a given year, and of those, as few as 10% may be pollinated. When not in bloom, these leafless epiphytes cling to swamp trees like green noodles, blending in so completely with their hosts that they are all but invisible. You might know epiphytes[25] by another name: air plants.

Air plants grow on top of other plants (typically, trees) by

---

[25] *Epiphyte* is translated from the Greek (*epi* = on top of; *phyte* = plant).

wrapping their roots around them. Sounds parasitic, I know, but it's not like that with epiphytes. They don't harm their host. They capture all that they need—water and nutrients—from the air. When they do bloom, their white, spectral blossoms seem to levitate in midair. That's owing to the fact that the flowers grow on a thin spike that reaches outward from its tree-clinging roots.

They smell of apples. Their sweet nighttime scent attracts . . . what, exactly? . . . Good question. In the center of each flower is a deep well of nectar. These ghost orchids have long nectar tubes, 13 centimeters (5 inches) or more in length. Given the length of the tube, it stands to reason that only something with a very long tongue could reach inside it. Just as Darwin predicted that a moth was the pollinator of an orchid with an even longer nectar tube and Wallace predicted that such a moth could be found on Madagascar, the scientific community just naturally *assumed* that the ghost orchid, too, was pollinated by a moth. And there's really only one moth with the "skills to pay the bills" for this type of job—the giant sphinx moth (*Cocytius antaeus*). With a proboscis twice the length of its body, its long tongue can access nectar in long-stemmed flowers. It makes sense, right? Of course. Except for the fact that no one had actually ever *seen* a ghost orchid be pollinated. It was just a guess.

Then in 2019, something remarkable happened. With the aid of camera traps and frequent visits, biologist and conservation scientist Peter Houlihan was joined by photographers Carlton Ward Jr. and Mac Stone on a quest to document the pollination of the ghost orchid. What they discovered upended everything we thought we knew about the ghost orchid. It's more complex than we suspected.

First, the team's images confirmed that the ghost orchid is *visited* by giant sphinx moths, but as it turns out, these moths had next to no ghost orchid pollen on them. Say what? Yeah. In their images, the sphinx moth can be seen drinking nectar, but its head is not nearly close enough to the flower to pick up any pollen. It's possible that the sphinx moth's association with the ghost orchid is completely one-sided.

Instead, the images documented *five* species of moths that interacted with the ghosts. Two of these species, fig sphinx (*Pachylia ficus*) and pawpaw sphinx moths (*Dolba hyloeus*), had ghost orchid pollen *on their heads.*

The team's groundbreaking work is important for two reasons: First, it's a reminder to us that we don't know something . . . until we have proof or data to back it up. Second, we had been laboring under the impression that there was only one thing capable of pollinating this rare orchid. Now we see that there are other moths that can do the job. This is such good news and makes the ghost orchid's plight ever so slightly less dire.

There's one thing that I hope you understand, though: This is important work, but I never said it was easy work. This team had to traverse dark, hard-to-reach, humid, stinky swamps. They shared space with bears, panthers, alligators, and several venomous snake species. More than 6,800 camera-trap hours produced a robust collection of images that had to be reviewed and analyzed. The team culled through more than 52,000 shots just to find a handful of illuminating images. It was time-consuming and painstaking work. We are all slightly better off as a result of their efforts and slightly more informed than we were before. You can see where I'm going with all this—we owe a lot to Houlihan, Ward, and Stone. We owe a lot to Dr. Walton.

Okay. That's all I'm going to say on the subject of moths. *For now.* Next I'd like to talk to you about flies. Nobody likes flies, do they? They're pests. They're hard to kill,[26] and the way they eat—barfing on things and sucking it back up again—is off-putting. But flies are actually butt kickers when it comes to pollination.

There are two types of fly pollination. Under the first kind, adult flies feed on nectar and pollen and regularly visit flowers. Under the second kind, adult flies regularly visit poop and dead things. We won't be talking about the second group just yet, although some plants do mimic stinky and smelly things to lure them.

More than 100 cultivated crops depend on flies for their pollination services—pears, apples, strawberries, cherries, plums, apricots, peaches, raspberries, blackberries, roses, mangoes, fennel, coriander, caraway, onions, parsley, carrots . . . The list goes on. A large number of wild food plants, medicinal plants, and cultivated garden plants are also aided by flies.

If they have their druthers, flies will seek out flowers that are dull or dark-colored. They like things that smell bad—poop, rot, blood, and the like.

Some flies, such as those in the family Syrphidae, masquerade as bees and wasps. You can tell them apart by their wings—flies have only one set of wings while bees and wasps have two. Of the flies that flock to flowers, hover flies lead the way. At 6,000 species

---

[26] So many cool studies on flies! They are hard to kill because they process sensory information so much faster than we do. We see things at a rate of 60 hertz. A house fly sees things at 250 hertz, meaning they take in more than four times the visual information per second than we do. Their rate of perception and ability to react to what they see is so much faster than ours, they can dodge a swat fairly easily.

worldwide, they are by far the most numerous of the pollinating flies and, perhaps, the most welcome. Not only do the adults pollinate flowers but the larvae also feed on aphids, key crop pests.

Hover flies are largely generalists, but some have unique relationships with orchids. The slipper orchid (family Cypripedioideae) uses its flashy yellow petals to draw in the hover flies, but its petals form a trap. Once inside, a hover fly will find nothing of interest—no nectar—but in order to escape the trap, it must drag its pollen-coated body through a small opening, where it inadvertently transfers the pollen and fertilizes the flower.

Another species of orchid, *Epipactis veratrifolia*, goes one step beyond that. It mimics the alarm pheromones of aphids. Hover flies flock to it in droves, eager for a quick and easy meal that, of course, does not exist.

POLLINATING BLOODSUCKERS

You probably don't want to talk about them, but mosquitoes are also flies. I know, I know: Mosquitoes are the worst. We're going to talk about how awful they are later on. But for now, I would be remiss if I didn't mention their role in pollination.

You have to remember that female mosquitoes are the ones that suck blood. They only need a blood meal right before they produce eggs. The protein in blood fuels that. The normal food of both male and female mosquitoes is nectar. They don't gather pollen like bees, but as they fly from flower to flower to feed, they pick up pollen and carry it from one blossom to another.

Mosquitoes aren't the hairiest of insects, but they can pick up pollen in unusual ways. The snow pool mosquito (*Aedes communis*), for example, eats the nectar from the blunt-leaved orchid, and its eyes often come in contact with the flower's pollen while it's foraging. The pollen sticks to its eyes. When it eats from another flower, the pollen touches the stigma of that flower, and the flower is pollinated. That said, mosquitoes are still pretty terrible, and we will trash them later.

MIGHTY MIDGES

Let's take a break for a minute, shall we? We've been working so hard on this. My head is full of insects, and I'm sure yours is too. Hmmm . . . I could really go for a little chocolate right about now. In times of stress or, in our case, mental fatigue, chocolate is very good for you.[27] And so delicious. Is it any wonder that the Latin name for the tree that yields chocolate—*Theobroma cacao*—literally translates to "food of the gods"? We should eat some chocolate immediately to maintain our mental acuity and stamina and resume the level of stress-free happiness that you were experiencing before you met me.

I probably eat too much chocolate. Americans consume 1.2 billion kilograms (2.8 billion pounds) of chocolate each year; that's more than 4.9 kilograms (11 pounds) per person. I'm sure I personally cover my share of that, if not more. The Swiss have us beat—8.7 kilograms (19.4 pounds) per person.

Did you know that it takes about 400 cacao beans to make one pound of chocolate? You need about 40 beans to make a Hershey's

---

[27] And healthy. Chocolate is derived from cacao beans. Everyone knows that beans are a vegetable. The cacao beans are mixed with sugarcane or sugar beets. Beets and cane are plants, which also puts them in the vegetable category. Which means, essentially, that Junior Mints are vegetables and therefore healthy. This is called deductive reasoning. You can see that I am exceptionally good at it. You must be so pleased that you have me as a guide on all this insect business.

bar.[28] A single cacao tree yields an average of 20–30 pods per year. Each pod has 20–40 beans.

For a bean to grow . . . Aw, you know where I'm going with this, don't you? There's no fooling you. I'm still talking about pollination, and I'm still talking about insects. Let me just throw this out there: Without *midges*, there would be no chocolate. Okay? Think about that. Would life even be worth living? Probably not. These tiny flies in the Ceratopogonidae and Cecidomyiidae families are the *only known pollinators* of the wee, white flowers of the cacao tree.

These downward-facing flowers can be found on the trunk and lowest branches of the tree. Midges are ordinarily attracted to mushrooms, but it turns out cacao flowers have somewhat of a mushroomy stink to them.

The native habitat of chocolate midges is a dense, shady rain forest. Today cacao trees are largely cultivated on open plantations— less to the midges' liking. As a result, these plantations often produce a lot of flowers, but on average, only three in 1,000 become pollinated and produce the much-coveted seedpods. Long odds, to be sure.

While I will tell you that these midges are the cacao tree's best hope at pollination, I will also tell you that the relationship between these poppy-seed-size insects and our beloved chocolate trees is not without its flaws. For starters, the midge's flying skills aren't worth much. It's quite an unreliable flier. Second, the dime-size blossoms of the cacao flower are a booger to get into. Each petal curves into a tiny hood. The hood fits around the plant's anthers, which the midge has to have access to. Third, between 100 and 250 grains of pollen are needed to fertilize 40–60 seeds.

---

[28] 1.55 ounces

Midges *might be able* to carry upward of 30 grains at a time—one midge can barely carry enough pollen to pollinate one flower. Fourth, that pollen is useless for any blooms on the tree it came from. The midge has to move to a different tree altogether for it to work. Fifth, the flowers have a narrow bloom window, typically lasting between 24 and 48 hours. Given all that we know, it's a wonder the midge has any success at all. It's just one more example of how complex the relationship between plants, insects, and people can be.

THE WORLD OF WASPS

We'd best get back on track here . . . Where were we? Ah, yes! No doubt you've been asking yourself this question for years: *What is the point of wasps?* It's a fair question. They are wretched little beasts. Yes, they do make artful, papier-mâché–like nests and that's cool, but getting stung by one downright smarts. Because they're carnivores, some people refer to them as meat bees.[29] So we hate them. They are hate-worthy. Except for one small thing: They are ecologically important little beasts. Sorry to ruin things for you.[30]

Because wasps are voracious predators, they tend to keep the numbers of potential crop and garden pests like aphids and some species of caterpillars in check. One study in the United Kingdom cited wasps hoovering 14 million kilograms (31 million pounds) of insect prey in a single summer. And while they are performing that public service, they are also inadvertently serving as pollinators. A wasp will visit a flower for only two reasons—one: to scope out potential prey, and two: to guzzle nectar for some quick energy

[29] Adult wasps don't eat the prey they kill—they feed it to their young.
[30] They are also devoted mothers and fiercely protective of their families.

while on the hunt. They are not especially hairy, so most of their pollination is limited and accidental. Unless we're talking about the fig wasp (family Chalcidoidea). Fig wasps are responsible for pollinating almost 1,000 species of figs, which is kind of a shocker. The only fig I'm familiar with is the Fig Newton. Which probably shocks the fig connoisseurs of the world, but there you have it.

As we happen to be on the subject of figs, though, read this next bit carefully because I'm going to teach you some cool new words. The first new word is *syconium*. That's a sort of hollow ball at the end of a stalk. All the fig plant's reproductive parts are stored inside this thing. Interesting, huh?

The fig plant gives off a unique smell that female fig wasps really dig. That's how this plant draws them in. When a female smells this, she's motivated to get inside the syconium. But how? There's only one opening at the top of the ball, and it's very small. To get to the heart of the syconium, she has to squeeze through a channel so narrow that she tears her wings and antennae off to get through. Talk about commitment!

Once there, she hopes to lay her eggs. Fig plants come in two varieties, though—male caprifigs and female edible figs. If a female fig wasp enters an edible fig, she won't be able to lay her eggs. The female flower parts include a long stylus, which will hinder her. With no wings or antennae and no way out, she will eventually die there, entombed. Which is troubling because it means an edible fig has at least one dead female wasp inside.[31]

If a female wasp enters a caprifig, she'll find male flower parts

---

[31] Sorry. For those of you who love a good Fig Newton, this might be a bit disturbing for you. Perhaps you could think of it sort of like Cracker Jack? With a prize inside? No. Perhaps not. It's not as bad as all that, though. An enzyme in the fig called ficin breaks down the lady wasp's carcass into protein. The fig basically digests the dead insect, making it a part of the resulting fruit. The crunchy bits in figs are seeds, not anatomical wasp parts. Better now?

that are perfectly shaped to hold her eggs. The eggs will grow into larvae, which will develop into male and female wasps. After hatching, the male wasps will spend the rest of their lives digging tunnels through the fig.[32] Then the females will exit the fig through these tunnels and fly off to find a new fig, carrying precious pollen with them, and the whole thing starts all over again. The fig plant and the fig wasp have a common goal. They both need to reproduce. This arrangement—called mutualism—works out rather nicely.

You might be thinking, *That's all well and good, but why do I need to know this? I don't eat figs.* You might not know it, but figs are wildly important. They are, in fact, a keystone species in many tropical ecosystems. A keystone species is one on which other species depend to the degree that if it were removed, the ecosystem would change drastically. Monkeys, birds, bats, and more than 1,200 other animal species eat figs. Fig trees stay in fruit year-round, including times when other trees are not producing fruit, so they become an essential food source.

Figs aren't the only thing in need of wasp pollination. There are almost 100 species of orchids that rely solely on wasps. In order to get the wasps' attention, these plants often resort to trickery: They mimic the way a female wasp looks or smells.

Smell can be a powerful lure. A handful of scientists from the University of Western Australia and the Australian National University spent a great deal of time trying to figure out the exact chemical composition of the pheromone the Arrowsmith spider orchid (*Caladenia crebra*) emits in order to lure a type of solitary

---

[32] If you're keeping track, you've figured out that the male wasp never leaves the plant it was born into. Which, I think, is a bit sad.

wasp called a thynnine wasp (*Campylothynnus flavopictus*). It is, apparently, a cocktail of sulphur-containing chemicals the wasp cannot resist. In fact, during the lab studies, they found that if the orchid's smell is stronger than an actual female's smell, the male wasp will choose the orchid. Once he arrives, he is essentially tricked by the orchid and comes away with only a little pollen to show for his efforts, which helps him none at all.

Australia's endangered hammer orchids go one step further. Yes, they smell alluring to male thynnine wasps. That goes without saying. What male thynnine wasp could possibly resist the smell of pyrazine? Whatever *that* is. But beyond that, each flower has a modified petal that bears a striking resemblance to a female thynnine wasp. How convenient for the orchid!

The female wasps are flightless and spend most of their time underground, but when they're feeling amorous, they climb the nearest stick and await their Prince Charming. The prince usually comes in the form of the very nearest male thynnine wasp. He zips by and plucks her from the stick, and the two mate while in the air. That's the goal, anyway, but in the case of the orchid, the male gets so addlebrained by the flower's smell that he fails to notice that the object of his affection is not a wasp at all.

You might be wondering why this orchid is called the hammer orchid. It just so happens that this special petal that he's trying to mate with is mounted on a hinge. When he grasps it and attempts to fly off with "her," his momentum and the hinge mechanism swing him upside down. He lands right onto the orchid's pollen-laden stigma. Extremely frustrating for the wasp, but a total win for the orchid.

Scent is a common trap plants reserve for pollinators. You've heard of *Amorphophallus titanum*, right? No, wait. Probably not. Let me try again. You've heard of the corpse flower, right? It is the largest unbranched inflorescence in the plant kingdom. That part isn't important right now. What is important is that this plant heats up and then smells bad. Really bad. Like, rotting flesh kind of bad.

I've never smelled a dead body before, so it's hard for me to put this in context for you. I made a list of objectionable smells from my own life to draw from, which was an interesting exercise. You should try it too. Here's what's on my list, from mildly offensive to super wretched:

Bad milk. There's nothing worse than the *thought* of bad milk. The smell is bad, but the sadness over losing the milk is worse. So much so that if you suspect it may have gone bad, you make everyone in the family smell it to confirm your suspicions. Because if it isn't really bad, then you can still drink it!

Cat barf. The dog loves to participate because no matter how bad it smells, he's still interested in eating it.

Boiled cabbage. I have memories of it from my youth, when my mom would occasionally make something called galumpkis, a Polish dish made of stuffed cabbage leaves and hate. It was a horrible, horrible smell. Cabbage should never be boiled. It is against the laws of nature, I believe.

My son Liam's unwashed fencing gear. Teen-boy sweat can create a powerful stench. Otherworldly in its terribleness.

Raw sewage. I mean, it kind of goes without saying, right? I had to crawl down into a Washington, DC, sewer for a job once (long story). I was wearing a biohazard suit but still. The smell defies description.

I think the worst smell I've ever encountered was when Chuck and I accidentally stumbled upon a tannery on the outskirts of a major city. Oh, man. That's a smell so bad, you actually feel it on your skin. Decades later, I can still conjure it up.

Matched against this handful of truly bad smells from my life, I'd have to rank corpse flower somewhere after Liam's fencing clothes and before the tannery. Yeah. That seems fair.

We know this 2.4-meter (8-foot) plant pretty well here in DC because the US Botanic Garden has one in its collection. It blooms every 10 years or so, and when it does, stand back. Of course, people flock to see and smell it. We feel strangely compelled, and we aren't alone in this. That stench is vital to the plant's survival. It draws dung and carrion beetles to it. The beetles go out of their minds when they smell it. They skip along happily, thinking they are coming to feast on a meal of Rotting Goodness. Instead, they come up empty but dusted in the plant's pollen, which they end up transferring to the next plant on their next hopeful excursion.

Beetles have a long history on our planet. They were probably some of Earth's original pollinators, and they remain prolific. Of the world's almost 350,000 flowering plants, it is thought that beetles pollinate close to 90% of them. Very few plants are *primarily* pollinated by beetles, but many plants do attract them, especially those that have large, cuplike flowers; those that are typically open during the day; those that have heavily scented blossoms; and those that have leathery or tough petals and leaves. Flower colors range from white and cream to pale green or even burgundy. Although beetles do have color vision, they rely primarily on their sense of smell to find flowers. They're looking for anything spicy, sweet, musky, or rotten.

Beetles certainly aren't known for their body hair and don't have any specialized structures for picking up pollen, but pollen does cling to them as they move from flower to flower. They hardly care. They are far more interested in chewing their way through petals and leaves—leaving small holes, bits of plant matter, and frass in their wake. Frass, you may not know, is insect poo, which explains why beetles are sometimes called mess and soil pollinators.

Some plants are fine with that. They are so desperate for a pollinator's help, they'll put up with anything. They'll also do anything. They might even be willing to kidnap a pollinator. Enter the Amazon water lily (*Victoria amazonica*). The water lily must survive in an extreme environment. The Amazon River floods and recedes as the seasons change. The water lily adjusts to its annual life cycle by growing creeping rootstalks and new leaves from seeds, flowering at high water levels, fruiting as the water recedes, and surviving low water levels as seeds.

Those seeds, however, must be pollinated. When the lily blooms, its white flowers emit a pleasant scent at dusk. The temperature within the flowers rises until it's higher than the air around it. The warmth and pleasing smell attract large scarab beetles (*Cyclocephala hardyi*). Later in the evening, after the beetles have arrived and are scarfing down flower parts, the petals close, trapping the beetles inside. By morning, the beetles are well-dusted in pollen, but the petals stay closed. By evening of the second day, the flowers turn red. They no longer produce their pleasing smell. The petals slowly unfold. The beetles, slightly dazed and confused over their lost weekend, stagger out only to fly off to the next bloom, where the pollen they now carry will rub off on the stigmas of other lilies and pollinate them.

As far as abductions go, it isn't a bad situation for beetles. Their curiosity is rewarded in the form of delicious plant parts to eat, a warm and safe place to spend the night, and a secure place for them to mate should they be so lucky to find a willing beetle partner. Both the lily and the beetles benefit—a classic mutualist relationship.

Oh, very interesting, you say. *What a nice story. But do they pollinate anything useful? Like crops?* Indeed, they do. They pollinate the pawpaw (*Asimina triloba*). I'm not making that up; it *is* a real thing. I'll have you know that the pawpaw is the largest fruit-bearing tree species native to North America. They were cultivated by Native Americans. And it was once a staple in the diet of European settlers before apples and other fruits were brought over for cultivation. Lewis and Clark relied on the pawpaw to get them through some tough culinary times. George Washington ate it for dessert. Even Thomas Jefferson planted pawpaw trees at his home in Virginia. I've never tasted one, but pawpaws are said to taste like a mixture of banana and mango and have the range of consistency from custard to an avocado, depending on how ripe they are. They are pollinated by beetles.

Maybe you've never heard of the pawpaw before, and you don't care. Maybe it's okay with you if the pawpaw doesn't survive. Fair enough. But trust me when I say that you do care about pollination. There's a lot of other stuff you *do* want to eat, right? Then you need insects for pollination. If you're now feeling especially anxious about pollinators and you need to take immediate action to try to support them, I totally understand. Just skip ahead to Chapter Ten. There's a lot you and I can do to help starting today. If you're steady for now and want to learn more about how insects further intersect with our food supply—and sometimes *become* our food—then proceed with me to the next chapter.

3.

INSECTS AS FOOD[1]

STRANGER DANGER. THAT'S something we're all supposed to avoid. Don't talk to strangers. Don't get into a car with a stranger. Things you are never supposed to do, and here I am, doing all those things and—worse—I'm doing them *with* my son!

I can explain. Liam and I are in Denver, Colorado, because he has qualified for the Junior Olympics in saber fencing. I could talk to you for a long time about this—how hard he has worked to get there, how proud I am of him, but none of that is important right now. What's important is that Denver is the location of Colorado's first and only edible insect farm. What luck, eh? Here I am, researching insects as a food source, and we land there. Yes, I did just use those words together: *insects* and *food source*. I know, I know: For those of us who aren't used to eating insects, it sounds *eeew gross*. Let's withhold our judgment and see how this plays out.

Because the Mile High City is at a high altitude, because Liam's body needs a day or so to acclimate before he competes, and because our stress and excitement levels are off the charts,

---

[1] DISCLAIMER: If you are allergic to shellfish or crustaceans, this chapter is not for you.

I think that we could use a distraction.[2] Because Liam is such a dutiful son, because he is also intellectually curious, and because he is *my* son and used to my ways, he doesn't even flinch when I suggest we should check this place out. I believe his exact words are "Okay . . . but I'm not eating anything there." Excellent! With such a strong show of support, I immediately contact Dr. Wendy Lu McGill, the president and CEO of Rocky Mountain Micro Ranch, and ask if we can tour her facilities when we reach town. She readily agrees.

"Just tell me the address, and we'll catch a ride over there," I say.

"No, no," she says, "it's kind of hard to find. I'll just pick you up."

Bright and early the next morning, she comes by our hotel. We hop into Dr. McGill's car and go tearing down the road to wherever it is we're going. I'm not really paying attention because I'm very eager to talk to her.

The bespectacled Dr. McGill has a kind face and a warm, welcoming manner. Her gentleness offsets some of the "ick factor" that is often associated with her business. Dr. McGill is a farmer of insects—mealworms, mostly, and sometimes crickets. It might seem an odd vocation if you didn't know much about Dr. McGill's background. She comes from a long career working with international aid agencies in sectors such as water and sanitation, human trafficking, and rural infrastructure development. Her doctoral studies led her to focus on agriculture and food security. Since 2015, she's been tending to her micro livestock and selling her company's products to some of Denver's local chefs as well as to curious eaters around the world.

Like many who are interested in entomophagy (that's the

[2] Don't judge! It seemed like a good idea at the time.

technical term for eating insects), Dr. McGill was driven to start her company after reading a 2013 report authored by the Food and Agriculture Organization (FAO) of the United Nations. In it, the FAO lays out a compelling argument for why insects could be a sustainable path toward future nutrition.

You should know that bug eating is nothing new, she tells us as we speed along. Insects have been a common part of the human diet for thousands of years. They currently form part of the traditional diets of at least two billion people. Insects are eaten on a regular basis in more than 164 countries.

Dr. McGill was intrigued by the sustainability and small footprint of insect farming and by the nutritional benefits of eating insects. A helping of crickets has as much calcium as milk, as much iron as spinach, and as much B12 as salmon, she says.

Farming insects is incredibly efficient, she says. You use less land and less water than in traditional farming. You have faster yields, and insects are a nutritious source of food. I'm just getting this all down when Dr. McGill abruptly stops the car. I look up. Wherever we are now looks . . . a little sketchy. The car is idling in front of a fenced-off lot at the end of a strip mall. Dr. McGill leaps out of the car.

"Wait right here," she says.

I look back at Liam, who says, "Uhh . . . Mom?" I shrug.

"Let's just see," I say.

I watch Dr. McGill through the windshield as she wrangles a large padlock. She pops the lock, hauls back the gate, hops back in the car, and drives us inside.

Before we can register our surroundings, Dr. McGill jumps out of the car again, closes the gate, and locks it behind us. Liam and I slowly get out of the car. The lot is barren, save for a 40-foot shipping container. Welcome to the farm.

A NEW KIND OF FARM

Dr. McGill sprints ahead of us to one end of the shipping container, where another padlock awaits. Conditions for the farm have to be exact, she tells us, so we need to brace ourselves. I *am* bracing myself, but not for what happens next. Dr. McGill finishes with the lock and throws open the door. A sickening heat wave rushes over us. We stand in the doorway, stunned. To ensure the health of the insects, Dr. McGill keeps this solar-powered shipping container at a constant temperature of 80°F with 80% humidity. It's February in Colorado, but you'd never know it once you cross that threshold.

Dr. McGill ushers us into a small antechamber that holds a refrigerator and a freezer. We push open the inner door and step deeper into the farm. It's here that we are met by The Smell. What *is* that? It's not foul, exactly. It's just . . . oppressive. Oppressively humid and a tad, well, buggy. After a few breaths, we acclimate but immediately begin peeling off our coats because it's so warm inside.

To our left, a long wall is lined with tall racks on wheels. Each rack holds maybe 14 trays. Each tray is dated by week. Dr. McGill pulls one tray open for us. It's filled with a sawdust-like material mixed with milled wheat or oats. The most obvious object in the tray is a carrot, lying on its side.

However, if you stare at the tray, you start to see movement. You can make out the shapes of tiny, golden-brown worms. This is a writhing mass of darkling beetle larvae, more commonly known as mealworms. They are about as long as the tip of your pinkie.

Dr. McGill closes the tray and pulls out another. And another. And another. With each tray, we see a slightly older collection of larger larvae until, eventually, she pulls open a tray of adult beetles.

And that's it. That's all Dr. McGill needs for her farm. There's no soil. There's no irrigation system. There isn't even running water in this place. There's no need for it. The carrots serve as a sufficient water source for the insects. They eat the milled wheat or oats.

Dr. McGill only needs six to eight weeks before her mealworms are ready for harvest. Each tray can yield several pounds of food. About a quarter of her insects are retained for her breeding stock. The rest are flash frozen and packaged. Some are shipped to local restaurants, and others are sent off to be made into products sold to the general public. Dr. McGill's operation is so efficient, she even processes the exuviae—exoskeleton castings from molts—and the frass into a product that can be used to fertilize your garden.

Liam and I move up and down the farm, opening trays and observing the different stages of life around us. Any nervousness we might have felt before has been replaced by sheer awe. How immaculate! How smart! How perfect! But we also have questions. Dr. McGill is calm and patient, answering the questions she has, no doubt, heard and answered thousands of times. Her knowledge of entomophagy runs deep. In short, Dr. McGill is a genius.

She sees me eyeing the freezer, though, just wondering what's *in there*. An earlier harvest, it turns out! Lining the freezer are bags and bags of frozen mealworms, ready to be taken to market. Several local restaurants—Linger, Comida, The Welsh Rabbit—have created signature dishes around her insects. Her products are popular among vegans, vegetarians, and people who follow a paleo diet. She seems to have a healthy clientele of people who just *like* eating insects too.

After our tour, Dr. McGill secures the farm by locking all the locks and safely returns us to our hotel. She no longer feels like a stranger—more like a trusted expert. Still, even after our fantastic tour, it's hard for Liam and me to imagine taking that leap. Insects as food?

TRY IT. YOU MIGHT LIKE IT.

Oh, I know what some of you may be thinking: *No way, no how. I would never eat a bug. How disgusting!* Let me stop you there and tell you that you probably ingest some insect matter every day without knowing it. Insect bits have a way of getting into our food during the normal cycle of growing and processing it, you see.

Lucky for us in America, the Food and Drug Administration (FDA) keeps track of such things. If you've got some spare time and a stern stomach, the agency maintains a detailed list that outlines our country's food defect thresholds. Defects, I should tell you, are the FDA's way of labeling poop, mold, and insect or rodent parts. It refers to insect defects as *filth*, which can mean whole insects, insect parts, or insect poop.

How much is too much? You and I would probably say that *any* amount is too much, but the FDA takes a different stand on these matters. According to the agency, "It is economically impractical to grow, harvest, or process raw products that are totally free of non-hazardous, naturally occurring, unavoidable defects." As long as the number of defects remain low, the FDA argues they probably won't present a health hazard. I feel safer already.

The list comes with a handy glossary of terms. Make sure you hit this first because by reading it, you'll be able to make some important distinctions. *Contamination*, for example, is an addition of foreign material to a food product. Examples of this would be dirt, hair, poop, insects, or mold. Contamination is not to be confused with *damage* to a food product, which shows evidence of pest habitation or feeding, such as tunneling, gnawing, or the presence of egg cases. Neither term should be confused with *infestation*, which is the presence of any live or dead life cycle stages of insects. Right. Good to have *that* made clear. Wouldn't want to mix any of those up in our heads. The FDA also keeps track of rot, mildew, and mold—all very distinct things—and things that are made rancid or are shriveled.

As a general rule, the FDA is not keen on having whole insects befoul our food, but it's a bit more lenient when it comes to heads, legs, and other dross. The level of allowable defects seems to vary considerably from one food to another.

I don't mean to editorialize here, but maybe think twice about apple butter, okay? I say that with great sadness because I happen to like apple butter. But the FDA allows for an average of *five* or fewer *whole insects* (not counting mites, aphids, thrips, or scale insects) per 100 grams. If 100 grams is hard to picture, think of it this way: That amount is a little less than the weight of a stick of butter.

Another thing I noticed about this list . . . We had talked earlier about figs and how they are pollinated, so I know you won't be shocked that insects (and we now know which ones!) can end up in fig products. What's interesting, though, is that the FDA's threshold for fig paste specifically mentions insect *heads*. Thirteen, if you're curious. Thirteen or more insect heads per 100 grams of fig paste is a no-go for the FDA. But . . . what about the *bodies*? Why set a specific limit just for their *heads*?

Maggots, I'm afraid, seem to be a concern for tomato products. Canned tomatoes allow for some variation, though. For every 500 grams, you can have 10 or fewer fly eggs; *or* five or fewer fly eggs with one maggot; *or* no more than two maggots total. Anything more than that would be a hard "no" from the FDA. Tomato juice and tomato paste have equally cringeworthy choices.

Mushrooms? Just don't unless you're okay with just under 75 mites per 100 grams of drained mushrooms. Frozen broccoli is right behind the mushrooms—up to 60 mites per 100 grams allowed there. The allowance combos for canned or frozen spinach are too lengthy and too bizarre to go into here. Honestly! Is nothing safe?

Raisins. Raisins are safe, right? They have to be. Well . . . golden raisins can contain 35 fruit fly eggs as well as 10 or more whole insects for every 8 ounces. Those little boxes that kids carry in their lunches? Those are 1 ounce each, so that's *potentially* 4 eggs and one whole insect in each box.

Even spices aren't safe. Try sprinkling a little black pepper into your pot of (maggoty) spaghetti sauce, and you might be taking in more than 40 insect fragments with every teaspoon. How often do you use ground cinnamon? I use it often, but maybe not so much now that I know that it can have up to 400 insect fragments per 50 grams. Ground or crushed oregano? Wow. That stuff can be chock-full of insect fragments. Need to know more? It's all on the list.[3]

---

[3] The Food and Drug Administration's *Food Defect Levels Handbook* can be found at fda.gov/food/ingredients-additives-gras-packaging-guidance-documents-regulatory-information/food-defect-levels-handbook.

Now, I'm sorry I ruined your day with all this unwanted information. I'm just trying to make the point that you do already—albeit largely unwillingly and unintentionally—consume insects. Whether you would be willing to try this *intentionally* is another matter. You might need more convincing.

MAKING THE CASE

Anecdotally, I've run across a couple of common arguments that defend the practice of bug eating. Usually, these are trotted out for the benefit of nonbelievers—those of us who are not used to the idea (i.e., Americans and most Europeans). There's the John the Baptist argument and the lobster argument.

There's a description in the Bible of John wandering the desert, and it says that "his meat was locusts and wild honey." So the thinking goes that if insect eating was good enough for John the Baptist, it should be good enough for you.

The lobster argument is linked to the idea that we eat what has been normalized for us. For example: Who ate the first lobster? Who looked at that thing—with its impenetrable exoskeleton, its twitchy antennae, and its clickety-clackety claws—and thought, *Mmmmm . . . that's some good eating, right there!* It's hard to fathom. Yet, someone did, someone else followed, and then lots of people followed after that. Lobster used to be so cheap and plentiful, it was relegated to feeding livestock and prisoners. As it became harder to get, it became an expensive food reserved for high society, but the lobster never changed. It's still a bottom-feeding scavenger with beady little eyes.

The lobster is a good reminder that food is subjective and good food is always open to interpretation. The rest of the world may snicker at Westerners for not being enlightened about eating insects. But Westerners were never socialized to view insects as food.

There are many data-driven arguments to support it. You need to look no further than that FAO report from 2013. It's long and detailed. I know you are very busy and probably don't have time to delve into it. That's okay. I'm happy to fill you in. You may come away with a better understanding of our global food crisis. I certainly did.

Let's start here: We are currently a global population of nearly eight billion people. This figure is expected to reach nine billion before 2050. While our world does technically produce enough food to feed its entire population, we don't do a good job of distributing food or preventing waste. Consequently, there are nearly one billion people who are chronically hungry. To meet our future food needs, our food production will need to almost double.[4]

Yet land is scarce. Currently about 70% of our agricultural land goes toward meat production—that includes both land for animals to graze upon and land for us to grow food for animals. Significantly expanding the area devoted to livestock may not be possible.

It's a bit hard to parse out, but here's a general sense of the math: To produce 1 kilogram (2.2 pounds) of beef, cows need roughly 200 square meters (656 square feet) of arable land. Pigs need 50 square meters (164 square feet) and chickens need 45 square meters (147 square feet). Since you'll obviously be raising

---

[4] Global demands for livestock products alone are expected to grow from 229 million tons in 2000 to 465 million tons by 2050.

more than a kilogram of each, you're going to need massive amounts of space to provide for your massive herds, passels, and flocks. By contrast, an orchestra of crickets only needs 15 square meters (49 square feet) of space to produce the same amount of meat.

Remember Dr. McGill's farm—it easily fit into a 12-meter (40-foot) storage container with room to spare. Insect don't need farmland. A single 75-liter (20-gallon) plastic tub can support a population of 1,000 or so crickets. They can be raised in almost any environment. Your farm could be your basement.

THE FUTURE OF FARMING?

We know that livestock can take up a lot of space, but they also kind of stink up the environment. Believe it or not, this has been studied extensively. In 2010, Arnold van Huis, Marcel Dicke, and several other Dutch scientists conducted a pivotal study on greenhouse gas emissions. They looked at the volume of methane and nitrous oxide linked to manure and gas produced by farm-raised cows and pigs. Then they compared this amount to the volume of gases produced by lab-raised mealworms, crickets, locusts, sun beetles, and cockroaches. It's no surprise that the insects grew more rapidly and gave off less gas than their mammalian counterparts.

You might well be thinking: *Why are you making me learn this?* You need to know because it's vitally important. A whopping 18% of greenhouse gas emissions come from gassy cows and pooping pigs. That's actually more than all of the world's cars, trucks, and motorcycles *combined.*

I don't have the space or time here to describe to you how those scientists went about studying the gas emissions of a locust, but you should look it up. It's pretty fascinating reading, and it will make you appreciate that you don't have to do what they did.

But you know *why* this is important, right? Methane and nitrous oxide are lightweight gases that float, float, float upward until they eventually enter Earth's upper atmosphere. These gases form a dense layer that allows sunlight to shine through but prevents heat from escaping. So Earth is gradually warming up, and as it does, things go haywire. Our weather patterns get weird. Our glaciers melt. Our sea levels rise. Intense heat waves spark severe droughts, which lead to food shortages. It's no wonder that our Dutch friends advocate so strongly for more insect farming and less cow and pig farming.

Here's something else: Insects are prized for their feed-conversion efficiency, which basically refers to an animal's capacity to convert food into body mass. Raising insects for food takes significantly fewer resources—in terms of food or water—than raising other livestock. Let's map it out. Say we want to produce 1 kilogram of protein, and our options are beef, pork, chicken, or cricket.

Just hear me out. To get your 1 kilogram of beef, you're going to have to feed Mr. Cow 10,000 grams of feed and 22,000 liters of water. Mr. Piggie needs less—only 5,000 grams of feed and 3,500 liters of water. Mr. Chicken needs far less: 2,500 grams of feed and 2,300 liters of water. Okay, so far Mr. Chicken is in the lead. But check out our cricket numbers. To raise 1 kilogram of cricket meat, we only need 1,700 grams of feed[5] and less than 1 liter of water![6] Mr. Cricket wins, hands down.

When processing your livestock, you know there will be some amount of waste. It's not all pure meat. You're going to be discarding bones and fat and other unmentionable, inedible bits. But even here, your cricket has the others beat. The proportion of livestock that is *not* edible after processing a cow is 45%, 35% for a chicken, and 30% for a pig. But only 20% of a cricket is inedible. Please don't ask me *which* parts of a cricket can't or shouldn't be eaten. I have no idea! Check with the FAO.

Finally, which of these four animals is the healthiest for you to eat? You know I'm going to say cricket. I *have* to say cricket. The numbers bear out. And not just crickets, I'm afraid, but insects in general.

---

[5] Another benefit here is that insects can be fed on organic waste streams. I don't want you to dwell too much on what this means but think manure, pig slurry, compost, and the like. Which actually is an argument against eating them, if you ask me. Knowing that my mealworms may have been chowing down on poop before they met their maker and ended up on my plate . . . But those in the know argue that this is a good thing because it adds value to biowaste. I don't know about that.
[6] I thought this was pretty interesting. The FAO reminds us that insects are cold-blooded. This means they don't need as much feed as animals like pigs and cows because they don't have to burn through energy just to maintain their body temperatures.

NUTRITIONALLY SPEAKING, A GOOD IDEA

If I were speaking as a nutritionist—which I am in no way qualified to do—I would say this to you: Insects are packed with protein, amino acids, beneficial fats, vitamins, minerals, and antioxidants, and they are a prebiotic fiber.

Reams of scientific papers have been written about insects as a food and their beneficial properties. It can be overwhelming. I had to study up a bit so I could pass on what I learned about the legitimate health benefits of eating insects. Ready?

Let's start with protein. Protein is an important building block for bones, muscles, cartilage, skin, and blood. It plays a crucial role in almost all biological processes because your body uses it to build and repair tissues.

Along with fat and carbohydrates, protein is a macronutrient, meaning your body needs relatively large amounts of it. Unlike fat and carbohydrates, your body doesn't store it, so there's no reservoir to draw from when you need it. Most meat-eaters turn to beef, pork, chicken, or fish as their main sources of protein. We can add insects to that list. Consider that 100 grams of beef contain 29 grams of protein. That same amount of cricket comes with almost as much protein—20 grams.

There's an insect source that's even higher in protein. In southern Africa—Botswana, Namibia, Zimbabwe, and South Africa—billions of saturniid moth caterpillars are collected every year. These caterpillars mainly feed on the mopane tree, so they're called mopane caterpillars. They're harvested by hand, pinched and squeezed to expel their guts, then boiled in salted water and sun-dried. Then they're taken to market, where their value exceeds that of beef.

These caterpillars are more than 60% protein and also contain valuable minerals such as calcium, zinc, and iron. They are eaten as snacks but also serve as the foundations for many meals. A common way to prepare them is to make stew—300 grams of dried mopane worms, one onion, two green peppers, six tomatoes, a tablespoon of curry powder, and 500 milliliters of water. Simmer in a pot and *voilà!*

I told you that our bodies also need amino acids—here's how these fit in. Amino acids bond together to make long chains. These chains are the proteins we just talked about. Scientists have discovered more than 500 amino acids, but only 20 are used to make proteins in the human body. Of those 20, only 9 are considered *essential.* I will list them here for you, and your eyes will immediately glaze over: histidine, isoleucine, leucine, lysine, methionine, phenylalanine, threonine, tryptophan, and valine. These are essential because the human body cannot manufacture them. They must be obtained through our food.

Food that has all nine essential amino acids are called complete proteins. Can you guess where I'm going with this? Go ahead. Guess which foods have all nine essential amino acids . . . Eggs, yes . . . and poultry, yes . . . and dairy, uh-huh . . . and INSECTS!

The reason insects are so chock-full of amino acids is because they need these same things for their own health. They seek out and consume foods that have amino acids. What might your go-to insects be for this? One study found especially high levels in bee brood (*Apis mellifera*), soldier termites (*Macrotermes bellicosus*), African palm weevil (*Rhynchophorus phoenicis*), and silkworm larvae and pupae (*Anaphe infracta*).

Fat is generally thought to be bad. However, all healthy bodies need a certain amount of it. Fat gives you energy, keeps your body warm and your skin healthy, helps you absorb vitamins from food, and produces hormones. There are different types of fat: monounsaturated, polyunsaturated, saturated, and trans fats. The difference between these types lies in their chemical structure and how they affect your body.

In general, unsaturated fats are good, and you can get them by eating plants and fish. Saturated and trans fats are less good for you, and you can find them in almost all animal sources as well as fried and baked goods like Cheetos and HoHos.

Edible insects are one of the few animal sources that contains higher unsaturated fats than saturated fats. You'll sometimes hear people talk about omega-3 or omega-6 fats.[7] Your body needs them for brain development, controlling inflammation, and blood clotting. As it happens, crickets have a perfect omega-3 and omega-6 balance. If crickets aren't your bag but you still need fats, look no further than larvae, worms, and caterpillars.

Australia's witchetty grubs are among the edible insect species with the highest fat content. These large, white, wood-eating larvae of the cossid moth (*Endoxyla leucomochla*) are a much sought-after part of the Aboriginal diet. I've never tasted them myself, but they've been compared to nutty scrambled eggs. Some folks eat them raw, biting off the bottom portion and discarding the head. Others chew the grub until only the skin remains, which you can spit out if it unnerves you. For those of us who like our meat cooked, the grubs can be roasted over a fire or sautéed on your stove with a little butter and garlic. Some people eat them over

---

[7] alpha-linolenic acid (omega-3 fatty acid) and linoleic acid (omega-6 fatty acid)

pasta. Not only are they high in fat content but the grubs are also very rich in oleic acid, which is an omega-9 monounsaturated fat.

Your body only needs vitamins and minerals in small quantities, which is why they're referred to as micronutrients. While the amounts needed may be small, you still need them. Without them, you can damage your immune system as well as your mental and physical development. Let me give you an example. Anemia is a condition in which your blood doesn't have enough healthy red blood cells, which leads to reduced oxygen flow to your organs. Having anemia can make you feel tired or weak. If left untreated, it can result in pregnancy complications, heart problems, or even death. Sounds bad, and it is. What can fix anemia? Getting enough iron in your diet.

Consider this: Of the 496 million nonpregnant women, 56 million pregnant women, and 305 million school-age children who develop anemia each year globally, half can blame their condition on iron deficiency. A food source high in iron could certainly help. Did you know that the iron you obtain by eating crickets is twice as accessible as the iron you gain by eating red meat?

This is really interesting. Researchers at King's College London and China's Ningbo University conducted an investigation of how much of different minerals can be found in commonly eaten insects like grasshoppers, crickets, and mealworms. In terms of raw concentrations of these minerals, none of the insects could really compare with beef. For iron, only crickets come close to cows.

But here's the thing: The amount of minerals *present* in a given food source isn't the same as the amount *available* for the people who eat it. See, our digestive systems can't always easily absorb nutrients from our food. This is a property known as bioavailability.

This team of researchers found two ways to gauge bioavailability. The first was solubility—how readily minerals from food dissolve in water. Solubility matters because it indicates how much of a mineral can be absorbed by our intestines. Across the board, insects beat sirloin beef, and crickets contained twice as much *available* iron as red meat.

The team's second method created an artificial stomach that replicated the conditions of your gut. Intestinal cells in the artificial stomach absorbed nutrients as it digested, which gave a measure of mineral uptake in living tissue. Here, buffalo worms led the way over sirloin beef, while mealworms and grasshoppers had comparable levels with red meat.

In a separate line of study, some researchers are pursuing the idea that insects like the palm weevil larvae could combat zinc deficiency—another serious public health concern for many poor children and women. The larvae contain a whopping 26.5 milligrams of zinc per 100 grams, compared to only 12.5 milligrams in beef.

These studies make compelling arguments for the insects-as-food concept, but they also illustrate the complexity of our global food issues. What we eat, who eats it, and how it's eaten all become important questions as we forge ahead in our overcrowded world. We can generally acknowledge that food resources are scarce *now*. Billions of people struggle with food insecurity *now*. We need to look for solutions that ensure nutritious and environmentally conscious food choices for everyone.

I don't want your brain to implode, but think about this: The nutritional profile of an insect can actually be altered depending on what you feed it. Need more vitamin A in your diet? Start feeding your crickets carrots. Okay, okay, I know: You'd rather eat the carrot. But what if carrots aren't available to you?

This idea has already been field-tested. When grasshoppers in Nigeria were fed with bran, which contains high levels of fatty acids, they had almost double the protein content of those fed on maize. You understand what this means, right? We could engineer a solution using insects to fight many of the health and food issues across the world.

As we continue assessing the health benefits of eating insects, we mustn't forget antioxidants. What even *is* an antioxidant? These things help protect your body from cell damage caused by free radicals. While this might sound vaguely political, free radicals are actually waste substances. They are produced as your body processes food and reacts to the environment. External stressors, such as pollution or cigarette smoke, can trigger their production. Oxidative stress on the body has been linked to a host of bad things—heart disease, cancer, arthritis, strokes, respiratory diseases, immune deficiency, emphysema, and Parkinson's disease. Antioxidants help neutralize free radicals and boost your overall health. Vegetables and fruits are rich sources of antioxidants. So are insects. Extracts of grasshoppers, silkworms, and crickets displayed the highest values of antioxidant—*fivefold higher* than fresh orange juice. I think I'd rather drink the orange juice than silkworm juice, but nutritionally speaking, I'd be wrong.

A healthy body needs carbohydrates too. Carbs can mean fiber, and in insect form, fiber means chitin. You know what chitin is—it's that stuff that makes up an insect's exoskeleton. Completely indigestible. As you're chomping down on your crispy crickets, you're getting all those powerfully useful nutritional benefits, but that exoskeleton is going down hard. *Your* body won't be able to break down *that* body, so it's going to have to pass through your system. Thus, it acts as a prebiotic. Prebiotics are types of dietary fiber that feed the friendly bacteria in your gut. I care deeply about the health of your gut, which is why I feel so compelled to talk to you about this. Chitin helps your gut bacteria produce nutrients for your colon cells and leads to a healthier digestive system. If you're looking for something to round out your diet, entomophagists would recommend you try the African migratory locust, which has a very high fiber content.

NO, REALLY. IT'S A GOOD IDEA.

I don't care! you scream. *I'm NOT eating a locust. Or a cricket. Or any of that stuff.* Fair enough. Calm down. I didn't say you had to eat one. I'm just building my case. My very compelling, irrefutable case that insects are a nutritious food source.

Are there any other arguments to be made in support of insect farming? We could talk about disease. All farm animals naturally carry a range of diseases, some of which can also affect humans. These are known as zoonoses. There's all sorts of nasty ones out there with regard to farming—*E. coli*, Salmonella, bird flu, and mad cow disease all come to mind. Yet insect anatomy is so different from human anatomy. Compared to mammals and birds, insects may pose less of a risk of transmitting zoonotic infections to humans.

We could also make a moral argument to support insect farming. Raising insects for food may be more humane than raising other animals for food. If you know anything about livestock husbandry systems—and no, I don't expect you to—you might have read the Brambell Report from 1965. The British government appointed a committee to look into the welfare of farm animals after a controversial book—*Animal Machines* by Ruth Harrison—was published. To a great deal of public outcry, Harrison's work illuminated the livestock and poultry farming practices of the time, which could best be described as grim.

The Brambell Report described the standards that the animal production industry should aspire to, which included freedom from hunger, thirst, discomfort, pain, injury, disease, fear, and distress as well as the expression of normal behavior. Britain's Farm Animal Welfare Committee (FAWC) developed these into the Five Freedoms,[8] which became an important framework for animal welfare from that point onward. We want animals to have room to roam. To live happy lives before . . . they are led to slaughter. Even with the framework in place, ensuring these freedoms for heavily farmed animals can prove difficult.

It's possibly a bit easier to do with insects. Providing food and water for them is less challenging than it might be for large mammals. For many insects, being reared in small spaces *is* part of their normal behavior, so that doesn't stress them out. Little is known about the extent to which insects can experience pain or discomfort. However, insect-killing methods like freezing may reduce suffering.

There's one final argument to be made in support of insect eating: Many people the world over *like* to eat insects. It's not a famine food for them—to be eaten only because there is nothing

---

[8] They're the freedom to stand up, lie down, turn around, groom themselves, and stretch their limbs.

else to eat. No. Many cultures have good associations with eating insects.

It's also a more common practice than you'd think. More than 1,900 species of insects have reportedly been used as food. Isn't that astounding? So many. Most of us get overwhelmed by too many choices. There was a time when M&M'S used to be plain or peanut. Today, there are many *other* kinds—almond, dark chocolate, mint dark chocolate, caramel, white chocolate, peanut butter, pretzel . . . Pretzel M&M'S, what the heck? Whose idea was *that*?

What was I saying? Oh yeah. That number: 1,900. You know me. I can't get past a number like that. Who says there are 1,900 bugs we can eat? How do they *know*? Who made the list? Did the person who made the list try each and every one? Did that person ever make a mistake and accidentally poison themselves? I had to know.

It was Yde. Yde Jongema. He's a taxonomist of the Laboratory of Entomology of Wageningen University, and he made the list—or compiled it from scores of scholarly citations, from the looks of it. The list itself is 100 pages long. It's a lot to take in.

Jongema tries to put it into perspective for us by helpfully grouping the most commonly consumed insects by type:

Beetles: 31%
Caterpillars: 18%
Bees, wasps, and ants: 14%
Grasshoppers, locusts, and crickets: 13%
Cicadas, leafhoppers, plant hoppers, scale insects, and true bugs: 10%
Termites: 3%

Dragonflies: 3%
Flies: 2%
Other orders: 5%

Wow. Okay. Beetles are at the top of the list. Now we know.

WHO'S HUNGRY?

Mexico, Brazil, Thailand, Japan, Ghana, and China are some of the countries where insect-eating is most widely practiced. Mexico certainly embraces the practice and has done so for centuries. Fried caterpillars, chocolate-covered locusts, and ant eggs soaked in butter are all favorites. Fried grasshoppers are very popular and are known as chapulines. Restaurants in Mexico City often include ant larvae, stink bugs, and water bug eggs on their menus.

In Brazil, içás (or queen ants) are a favorite snack. In Colombia, big ass ants are a delicacy. No, I'm not trying to be rude. That's the literal translations for *Hormigas culonas*, which are collected in the Santander region.

In Thailand, fried bugs are commonly served as snacks. One of the country's most popular ones is jing leed, a deep-fried cricket seasoned with Golden Mountain sauce (similar to soy sauce) and pepper. Other favorites include grasshopper, woodworm, bamboo worm, and mang da, which is a really big water beetle.

Restaurants all over Japan serve up hearty portions of hachinoko (boiled wasp larvae), sangi (fried silk moth pupae), and zazamushi (aquatic insect larvae). Fully grown insects such as semi (fried cicada) and inago (fried grasshopper) are also eaten.

In Ghana, termites are a snack and, at times, a necessity. Other types of food are often in short supply during the country's spring months, when many Ghanaians are busy planting crops. Luckily, the season's heavy rains force winged termites to flee their underground homes. The termites are high in protein and fats. They can be fried, roasted, or ground into flour. Insects can account for up to 60% of the dietary protein in a rural African diet.

People in China snack on water bugs boiled in vinegar, roasted bee larvae, and fried silkworm moth larvae—all of which are rich in nutrients like copper, iron, riboflavin, thiamine, and zinc. And when temperatures begin to drop, they keep warm with a steaming bowl of ant soup.

As you can see, a lot of people around the world eat insects and happily so. Knowing all that, I still can't envision myself eating any on purpose unless I was in a dire situation.

MORE THAN JUST SURVIVAL

Say we're on a survival show. You and I have to live off the land for 21 days by ourselves with only a fire starter and a machete. Except that I've got my little burlap bag stuffed with trail mix and Junior Mints. So I'll probably be okay. But you, you are going to have to eat nasty stuff and drink water from scummy mud holes. Because that's sort of the point of survival shows.

Do we have any survival skills between the two of us? I can't speak for you, but I will definitely not be of any use to you whatsoever.

I was pondering this very scenario one day when I came across an interesting website run by a Missouri-based survival training company called Sigma 3 Survival School. On their website, there were two videos, first on how to hunt grasshoppers in the wild and second on how to prepare them and eat them. Grasshoppers.

Now, see, I've got a thing about grasshoppers. It's a pathological fear that's probably ripe for psychoanalysis. I could no more EAT a grasshopper than I could stand to watch these videos.

In the first video, Rob Allen—founder, president, and head instructor of Sigma 3 Survival School—has cut down a small, leafy branch from a tree. This is his "stunna stick," which he will use to stun or immobilize grasshoppers in the big field behind him.

I watched transfixed, almost paralyzed. I can feel my heart rate go up. No, Rob. No. Don't go in there. Not the grasshopper-infested field! He's going. He's going into that field, and it's probably teeming with hundreds of grasshoppers . . . hiding among the grasses . . . lying in wait. Good grief. Rob is so brave. I swallow hard.

I probably shouldn't worry. Rob is an able survivalist. He's been an outdoorsman all his life. He's an Army veteran who served in Iraq. He's skilled in field infantry tactics and has had combat trauma training. Rob started his survival school 10 years or so ago not just to help people learn to survive in an emergency situation but to become fully self-reliant in the wilderness. He hopes to instill an awareness in people so they become better stewards of our planet. Rob has seen some stuff, I'm sure. And clearly, he's not about to get his briefs in a bunch over a few grasshoppers. Not like some of us.

As he walks along in the video, Rob is smacking the ground with the tree branch, knocking the daylights out of grasshoppers.

"As Americans, bugs are kind of repulsive to us," he says. "But the majority of the world eats them on a daily or regular basis.

In a survival situation, this is a huge consideration for you. Especially in the beginning. Because, you know, we have our survival priorities—we take care of our shelter, our water, our fire, our food. You can use grasshoppers as a supplement to your diet or even as a mainstay." *Hmmmm . . . not likely, Rob,* I'm thinking. As he wields his stunner, Rob picks dazed grasshoppers off the ground and pops them into a plastic container. I can hardly breathe when he holds up the container for viewers to see.

In the next video, Rob is sitting by a campfire. He has a long stick, which he is stripping the bark from. Bark, he is explaining, is full of tannins.

"I'm removing the bark because a lot of barks have tannins in them, and we don't want any kind of wood that's going to impart a nasty flavor . . . so we're just going to take the bark off this so we don't get any unwanted flavors in our grasshoppers," he says. He must be joking, right? Tannin is an astringent. Astringents pucker the mouth, numb the tongue, and constrict the throat. But that *has* to be better than actually tasting a grasshopper, doesn't it? I would need my stick to have *extra* tannins.

Next he makes a long slit down the center of the stick. He's creating a sort of clamp to poke the grasshoppers through so he can roast them over the fire. The plastic container sits nearby. He reaches in it and pulls out a long, hefty grasshopper.

"One of the most important things about eating grasshoppers," he tells us, "is we remember to remove the legs and the wings. Especially the big back legs. Those will get caught in your throat."

Rob deftly plucks off the grasshopper's legs and wings. Then he slips its body into the cleaved stick. As he sets the stick over the fire to roast, he says, "These grasshoppers are a complete food, guys.

I mean, you can literally survive on these things. It's definitely a food source that we take advantage of in the summer."

Later in the video, after his stick o' grasshoppers has fully cooked, he pulls one off and eats it. I think I must have passed out for a brief second after he did that. I'm not sure.

As cowardly as I am, I feel the need to speak to Rob directly. To gain insight. He is so obviously a courageous and righteous soul; perhaps I can learn from him.

In our first conversation, I aim to establish a baseline with Rob about what sorts of things he has actually eaten during survival situations. It turns out, the list is pretty extensive. He eats small game and fish, sure, but he also eats some of Nature's more interesting fare: raccoon, squirrel, skunk, opossum,[9] mice, and rats.

---

[9] "As far as eating it in the bush, in a survival situation . . . it's awful. It's got a real dirty flavor to it . . . Any kind of scavenger—skunk, opossum, raccoon—they have a certain kind of flavor that's really unique to them." Yes. Well. One can imagine.

"You eat mice and rats?" I ask.

"Rats preferably just because they're larger," he says. "But a city rat is really different from a woods rat. A rat in the wild actually has a really clean diet because they're eating seeds and nuts, so they're pretty clean-tasting. But a city rat—I probably wouldn't touch a city rat unless I was just desperate."

Right. No. Me neither.

"And insects," I prompt. "You eat insects."

"As far as bugs are concerned, I try to keep it to arthropods—grasshoppers, crickets . . . They're very high protein," he says. "Probably my favorite thing to eat from the bug world would be grubs. Grubs are actually extraordinarily protein-and-fat dense. They're, like, 40% protein and 40% fat."

Good to know.

"So if you can dig in a log and eat one of those . . . if you can bake it or preferably fry it in the fat from some other animal like racoon or opossum, it tastes kind of like—a little bit like french fries."

Thank you for ruining french fries for me, Rob. But speaking of taste . . . what does a grasshopper taste like? It can't be good.

"They're not great," Rob admits. "They need some kind of seasoning to them. They're really plain-tasting. They're crunchy and kind of cardboard-like . . . very lacking in flavor in general."

"Is there anything that would make them taste better?" I wonder aloud.

"I mean, if you had some cayenne pepper or honey. You know, John the Baptist[10] ate locusts and honey, so those are options. But in a survival setting, not so much. There aren't a ton of options."

_____

[10] What *is* it with John the Baptist? He's becoming the poster child for locust eating.

By now, I've read all the statistics on what a great and healthy food source insects can be, which Rob also espouses. Still. It's hard to get past the idea of it.

"We can't continue to grow as the population has been and keep eating chickens and cows," Rob tells me gently. "There's a finite amount of resources for that."

"I know, Rob."

"At the end of the day, it's just food."

"I . . . I know," I mumble. "It's just . . . I just . . . I don't think I could do it. Eat a grasshopper."

"Yeah, you could. We do a seven-day program where people live off the land. I guarantee that by day five, you'll be ready to put anything in your face."

"Rob, by day five I think that I'd be dead."

"Nah, you wouldn't. I bet you'd survive a heck of a lot longer than you'd think. Women are mentally tough."

Yes, that's true. Women are. Not *this* particular woman. But *other* women. Wonder Woman could probably do it. Madeleine Albright could have probably done it. Meryl Streep could *portray* a woman who could convincingly do it. My mom could probably do it. They're all steely women. I hate that I'm not tough like that.

"I just don't think I could," I tell him miserably.

There's a slight pause as he considers this.

"Honestly, there's nothing wrong with fasting," he offers. "A person can fast for weeks and weeks and be just fine."

True, true. In a survival situation, I could just opt to starve and die.

But Rob, being the fine instructor that he is, isn't one to give up so easily. He sees another way for someone like me.

"You can always take those same grasshoppers that you got with the stunna stick and utilize them to catch fish or use them for primitive traps, such as deadfalls, to try to catch other, larger animals."

I perk up. Now you're talking! Bait!

"Everything that you do with survival is about efficiency; it's about taking the resources that you have and producing the greatest amount of calories for the least amount of expenditures," Rob counsels me. "If I can take a grasshopper and stick it on a hook and catch a large bass or a catfish, that's going to be a better use of my resources than sitting around trying to eat grasshoppers every day, all day. That's not really sustainable."

Amen to that, brother! Not sustainable and definitely repulsive to me. This is what I love about Rob. I know he will help me survive without asking me to go against my nature, which is to naturally revile grasshoppers. I feel slightly better about the whole thing. I would still have to technically *touch* a grasshopper after I stunned it and put it on a hook or something. But a fish would love to eat it, and I would love to eat a fish. There you go! Circle of life!

While all this is profoundly useful, I don't plan on being in a survival situation anytime soon. At least not without Rob. So I need to ask myself: Could there be any other circumstances where I might eat—if not grasshoppers—any sort of insect?

TRY IT. YOU MIGHT LIKE IT.

I know myself pretty well; if I'm going to try this, I'm going to have to *ease* into it. And if I'm going to do this, I need to be transparent with my family. I've heard stories about celebrities who grind up vegetables and slip them into their kids' brownies to trick them into eating healthy. I can't imagine a more ridiculous deceit. I'm certainly not about to trick my family into eating anything. If I make something with insects and they *want* to try it, they are welcome to. I would never force them, and I certainly wouldn't deceive them. If we're doing this, it's with our eyes wide open.

I search the internet for ideas and come across a company called Chirps. They market a cookie mix that uses cricket flour. I bring it up casually one night to the boys. Liam shrugs. How bad can it be? Devin, my youngest son, gives me a look. Chuck says nothing.

I figure that's enough of a green light. That said, when the Chirps bag of chocolate chip cricket cookie mix arrives, I approach it with all the care you might use to handle live munitions. *It's just cricket flour,* I tell myself. *It's not like you're going to open the pouch and see little cricket heads with little cricket eyes staring up at you accusatorily. These crickets have been ground up into a fine powder.*

What if it smells? my cowardly self inwardly whimpers.

Are crickets known for their bad smell? My small-but-courageous-self thunders back. *Grow a spine, girl!*

I gingerly snip open the bag and look in. It looks like cookie mix. I take a whiff. It smells like brown sugar and cinnamon and oats. Did I make a mistake? I check the back of the package again. No, it definitely says cricket all over the packaging. All the ingredients are listed:

Wheat flour
Oat flour
Cricket flour
Baking soda
Salt

Semi-sweet chocolate chips

Granulated sugar

Brown sugar

I'm supposed to add some butter, one egg, and a bit of vanilla extract. Well, that doesn't seem so terrible. It's just like regular cookies, really. Except for the cricket part.

According to the packaging, cricket protein comes with a deliciously nutty flavor. It has 5 grams of protein, more B12 than salmon, more iron than spinach, and all nine essential amino acids. It's also a good source of zinc and is soy free. Each cookie I make will have the equivalent of 20 crickets.

There are two types of crickets that make up this particular flour. I don't have to wonder about this because they are helpfully listed on the package: *Gryllus assimilis* and *Acheta domesticus*. If you don't know to look, your eyes would probably have gone right over those ingredients. You certainly wouldn't be prompted to google them as I did, and you wouldn't immediately be skeeved out to learn that *Gryllus assimilis* is Latin for Jamaican field cricket, and *Acheta domesticus* is Latin for the common house cricket. And you certainly, *certainly* would not then do a Google search for images to see their little crickety faces. No, no, you wouldn't do any of that, and I strongly recommend that you do not. Too late for me, of course, but I can spare you that much.

Matters are made worse, of course, after I read little Google profiles about them. The Jamaican field cricket has been falsely dubbed the silent cricket, for reasons unknown. It definitely chirps. It chirps at a pulse rate so high with intervals so brief that it comes out sounding like a continuous sound instead of individual chirps. I'm not comforted to know that both are considered pests or that both are largely bred in captivity for the pet industry. So all you reptile lovers out there are feeding your snakes with these things.

I'm filled with unease.

Stop it! Do your part for the planet! I tell myself firmly. *Get in that kitchen and bake those buggy cookies!*

Before long, I have them in the oven. The smell of chocolate chip cookies begins wafting through the house. Liam is the first to appear in the kitchen.

"Whatcha got there, Mom?" he asks.

"These are the cricket cookies I told you about, Liam."

"Uh-huh."

"So, you know, you might not want to eat them."

"They smell good," he says.

"Right. They do. But they have ground-up crickets in them, Liam."

"So you said. I'm sure they'll be fine!"

As I take them out of the oven, Devin appears.

"Are these your bug cookies, Mom?" he asks skeptically.

"Crickets, yes," I say. "So you might not want to eat any. I mean, you don't have to. At all."

The three of us stand staring at the tray of cookies while they cool. They look surprisingly normal. They smell really good.

"It's just an experiment," I say. "You know. To see what it's like."

"I'm sure it'll be fine," Liam reassures me again.

"Do you want me to go first?" I ask. But before I can, they both pick up cookies and bite into them.

"They're good!" Liam says.

Devin nods. "They are."

I taste one. It's actually quite delicious.

"They're very rich," Liam says. "That's because of the added protein from the crickets."

He isn't wrong.

Chuck wanders in. "Oh! Cookies!"

"Crickets, Dad," says Devin.

"Oh . . . um . . . maybe later," Chuck mutters.

"Dad, they're really pretty good," Liam says.

"I'll take a pass," says Chuck.

I believe this first attempt is a success! Before long, I have purchased my next experiment: cricket brownie mix from a company called Cricket Flours. This time, I am to add water, oil, and eggs to a normal-looking brownie mix made with *Gryllodes sigillatus*, the Indian house cricket. It is by far more disgusting looking than the other two, in my opinion, and I am once again made grateful that they have been grounded up into dust so that I don't have to actually see any of them. The entire pan of brownies is equivalent to 150 crickets.

The boys taste the brownies and declare them basically normal, although we all agree they are richer than our normal brownies, "on account of all that extra protein," Liam says.

"No, thank you," says Chuck, once again.

My next purchase is a bag of cricket flour, which can be used in all sorts of ways. You can buy this type of flour and all sorts of bulk edible insects. If you're interested, there's a whole genre of cookbooks available that honor the insect. The famous London eatery Eat Grub recently published their *Ultimate Insect Cookbook*.

It has a recipe for mealworm bread that looks palatable and a spiced grasshopper, butter bean, and lime hummus that might be tasty. Arnold van Huis, Henk van Gurp, and Marcel Dicke's *The Insect Cookbook: Food for a Sustainable Planet* has that mopane caterpillar stew recipe among many other culinary delights. Or you can read through David George Gordon's classic *The Eat-A-Bug Cookbook*.

SO MANY OPTIONS

I'm not sure I want to make a habit out of this, but it does prompt me to start looking for edible insect products. You wouldn't believe how many are out there. Cricket Flours also sells cricket bites. You can get the five-pack gift of assorted flavors, which comes with roasted original, spicy cayenne pepper, hickory smoked peppery bacon, cheesy ranch, and buffalo wing sauce. These are actual crickets with cricket bodies and cricket heads and cricket eyes and cricket legs. I like hickory smoked peppery bacon as much as the next person, but I'm not sure Cricket Flours could coat them enough in spices to obliterate the cricket. So maybe I can't do that.

Hotlix makes insect lollipops with real bugs in assorted colors and flavors. They are kind of see-through, so you can definitely see the bug you're about to eat. Ant lollipops come in blueberry, apple, banana, or watermelon. Cricket lollies come in blueberry, grape, orange, or strawberry.

Don Bugito, which makes planet-friendly edible insect protein snacks, has love bug boxes that offer sweet and savory options: coconut brittle bugitos, chili lime crickets, granola with cricket flour, and spicy bugitos OR chocolate-covered crickets. Yum, yum! Meat Maniac offers BBQ mixed bugs mix—a whole bag of fun filled with crickets, sago worms, mole crickets, and silkworms.

If I wanted a quick pick-me-up, I could try the Exo cricket energy bar. Gosh. I hardly know what to do. There really are so many products to try.

Then I remember our friend Dr. McGill. I remember her loving care of the mealworms at her farm. Her products are called Insectables ("Snacks with a mission!"). I ordered a packet of roasted mealworms (sea salt and cracked pepper) and roasted crickets (Mexican spiced). She also sells cricket (flour) pasta and chirpy jerky (insect-based pemmican-style jerky made with crickets).

They soon arrive in the mail, but I feel significantly less brave when I eyeball the crickets. Their stiff little bodies are coated in a dark red spicy powder, which does not make them look any more appetizing to me. Baby steps. Maybe I should try the mealworms first? I open the pouch and pour a little pile of them into my hand. Touching them isn't the problem. They feel sort of dry and husky.

I pop the flaky things in my mouth and chew, grateful that I'm not feasting on anything more substantial. No witchetty grubs, no winged termites, no locusts. If someone offered me a locust to eat, I might faint. The mealworms taste . . . well, salty, really. Not much more than that. I take another glance at the cricket packet. Noooooooo . . . I don't feel quite ready for those. In truth, I might never be ready for those. You might not, either. You might, in fact, be more determined than ever to avoid eating insects. That's okay. I understand; I really do. But people aren't the only ones with an interest in eating insects. Plenty of animals rely on insects, and there are even some plants that eat insects.

ON EVERYONE'S MENU

We know that insects eat insects. Good grief—we have so many examples of that: ladybugs killing aphids; dragonflies killing mosquitoes (good riddance); mantises killing . . . everything. We know that birds eat insects. Reptiles and amphibians too. Fish. Lots of mammals eat insects—bats, hedgehogs, shrews, aardvarks. Many of the animals that eat insects are specialized to do so. Birds have beaks that can pierce through exoskeletons. Reptiles have sticky tongues that can snag their quick-moving targets. Some animals are uniquely suited to hunt and eat them.

Let's start with anteaters. That's an easy one, right? No mystery there. *It's in the name.* They eat ants. There are four types of anteaters from the species of the suborder Vermilingua (meaning worm tongue). Since anteaters have no teeth, they must use their long tongues to collect their meals. An anteater's tongue can flick up to 150 times a minute, helping them to lap up thousands of ants and termites in a day.

Not every animal that eats insects has so obvious a name. There are others—like the sugar glider, the sun bear, and the star-nosed mole—that you might not realize are insectivores. There are also animals that you may not have ever heard of that are insectivores. I'm referring, of course, to the moonrat, the numbat, and the aardwolf. Ha ha, you think I'm making those last few up, don't you? I wish! Those are actual animals that actually eat insects.

Let's go back to sugar gliders (*Petaurus breviceps*)[11] for a minute. If we are following the anteater naming convention, these little dudes must eat sugar, right? They do. They feed on nectar, pollen, acacia, and eucalyptus tree sap. They burn a lot of energy to

---

[11] This name translates to short-headed rope-dancer. Just so you know.

support their "flight." Looking like a cross between a dwarf hamster and a flying squirrel, these huge-eyed marsupials have a soft membrane between their wrists and ankles that allows them to glide from tree to tree. They can float up to a distance nearly the width of a football field. When they're not supporting their sweet tooth, sugar gliders chow down on moths, beetles, and crickets; and they would not turn down a mealworm or two, if offered.

I can picture an itty-bitty animal like a sugar glider eating insects, but a sun bear? What do I know of *Helarctos malayanus*? Next to nothing. It's the smallest, second-rarest, and least-studied bear from Southeast Asia. People always dither on about that cool pattern on the sun bear's chest that looks like a rising or setting sun (depending on if you are a glass-half-empty or glass-half-full kinda person). But forget the chest hair. With the sun bear, it's all about the tongue. Its tongue can be more than 20 centimeters (7.9 inches) long. The sun bear uses its fierce, curved claws to rip open trees in search of insects, then licks them up. Sun bears primarily feed on termites, ants, and beetle larvae. They also eat some kinds of fruit, with a special fondness for figs.[12]

How strange is that, you might be thinking. *A bear that eats bugs.* Not so strange, I guess, considering that the sloth bear (*Melursus ursinus*) is even more of a bug lover than the sun bear. Sloth bears, which can be found in the lowland forests of India and Sri Lanka, are like a shaggier, more disheveled cousin to the sun bear. This bear is also built for bugs.

An adult male's weight can top 140 kilograms (310 pounds) and, frankly, it's hard to imagine all that weight comes from insects.

---

[12] Do you think it knows about the fig wasps?

But in fact, that is the case. During non-fruiting season, insects make up 95% of a sloth bear's diet.

Sloth bears are expert hunters of termites and ants, which they track by smell and can locate underground. Once a sloth bear finds a termite mound, it paws it with scythe-like claws until it reaches the large combs underground. It blows away soil with violent puffs of air. Sloth bear snouts, lips, and teeth are adapted for insect eating. Their long lower lips can be stretched over the outer edge of their noses, and they are missing their front teeth, which clears the path for insect sucking.

The sloth bear sucks up the termites through its muzzle—a sound reported to be so loud and distinctive, it can be heard 180 meters (almost 200 yards) away. These bears are also fond of honey. Mother bears regurgitate a mixture of half-digested jack fruit, wood apples, and honeycomb. This sticky mess hardens into a dark yellow mass, which she feeds to her cubs. I'm not sure I'd be up for eating bear barf, but it is considered a delicacy by some.

As we know, nature is not always pretty, and not to be uncharitable, but the star-nosed mole (*Condylura cristata*) is some kind of serious ugly. All but blind, this hamster-size critter plows through soggy soils bobbing its head up and down, side to side, hunting for worm or insect prey. To the mole, timing is everything. The starlike structure on its face can touch a dozen different places in a single second. With each touch, 100,000 nerve fibers send information to the mole's brain. That's five times more touch sensors than in the human hand all packed into a nose smaller than a fingertip.

When the star touches something, the mole makes a quick decision about whether it is food or not, usually in about 8 milliseconds.

This mole can find, identify, and scarf down food in an average of just 227 milliseconds—less than quarter of a second.[13] For that, the star-nosed mole holds the special distinction of being the world's fastest eater. This mole is a serious eater and not to be trifled with. While it loves worms the most, about a third of its diet is the larvae of caddis flies, midges, dragonflies and damselflies, crane flies, horse flies, predacious diving beetles, and stone flies.

The reason I'm bringing these animals up—and it is a strange list, I know—is because you probably don't spend a lot of time thinking about them. Yet each of them relies quite heavily on insect diets. There are a few other animals I want to talk to you about, and these are animals that I honestly had never heard of. I didn't know they existed, but they do, and what they have in common is that they eat insects.

Our first stop on our tour of the bizarre is the moonrat (*Echinosorex gymnura*). This southeast Asian animal is neither rat nor rodent. It is a small, carnivorous animal related to the hedgehog, and it smells very, very bad. The fact that it reeks of ammonia is one of its defining characteristics. Another defining characteristic is that it eats bugs. Lots of them. By day, it snoozes under logs or in abandoned burrows. By night, it skulks around looking for arthropods and other insects to gobble up.

Next up: the numbat (*Myrmecobius fasciatus*). It's a marsupial native to Western Australia and eats termites. Almost exclusively. The numbat is a rather attractive little thing, if I'm allowed to say. It has sort of a squirrel head with a pointed nose. Its fur is banded, and it has a bottlebrush tail.

---

[13] This is fast. By comparison, it takes a person 650 milliseconds to brake after seeing a traffic light turn red.

Numbats dig small holes in the ground to uncover the passage-ways that termites travel in when they go to and from their nest. Termites are no match for the numbat's 10-centimeter (3.9-inch) saliva-coated tongue. Ridges in the soft palate of the numbat's mouth help it to scrape the termites off its tongue so it can swallow them. This 478-gram (1-pound) animal can knock back 20,000 termites in a single day.

Thankfully, the numbat and the aardwolf live on separate continents. Otherwise, there might be a daily turf war over termites. The aardwolf (*Proteles cristata*) is an insectivorous mammal native to East Africa. It is from the same family as hyenas. So picture a hyena that eats bugs. It's not at all what I picture when I think of hyenas . . . I always sort of picture them gacking down antelope entrails left over from a sated lion's bloody kill. But no, the aard-wolf isn't into entrails. It covets termites and snags them with its broad, sticky tongue. Being a much larger animal than the num-bat, it unsurprisingly has a larger capacity—to the tune of 250,000 termites a day.

Have you ever heard of these creatures? No? How is it that we've been on this great Earth of ours this long and things like this are living their busy little lives, and we have no idea they even exist? I find this mind-boggling. It's what I've talked to you about before: We don't know what we don't know. If insects are in decline in the areas where these animals live, would they adapt? Would they die off? I can't answer that for you, seeing as how I'm only just now getting up to speed about them being alive in the first place. It does make me wonder.

You know what else makes me wonder? Carnivorous plants. We don't generally think about plants needing to eat. Carnivorous plants eat. They lure, capture, kill, and digest animals. Mostly insects, but some of the bigger plants eat lizards, mice, or rats too. They do this not because they are sinister or bloodthirsty like Audrey II from *Little Shop of Horrors*, but because they usually live in areas where the soil quality is poor and they need to supplement their diets.

Most people are familiar with Venus flytraps (*Dionaea muscipula*). You might have also heard of pitcher plants. The meat-eating plant world goes well beyond these two examples, however. There are some 600 carnivorous plants around the world. Many are found in the hot and humid tropics, but some, like the Venus flytrap, can be found in the US.

There just so happens to be a place in Maryland not far from where I live called the Carnivorous Plant Nursery, run by Michael Szesze, a retired science teacher. Szesze sells many types of carnivorous plants, but for beginners like me, he sells a starter pack, which includes one Venus flytrap, one sundew, and one pitcher plant. Exciting! These little devils arrived through the mail wrapped in wet paper towels. I quickly potted them in special soil and waited for something interesting to happen.

Carnivorous plants work in wondrous ways, but they do not work on cue. Sitting around waiting for them to suck something up is not as thrilling and immediate as you might hope. I didn't want to give any of my new plants anxiety, so instead of gawking at them, I went inside to do some research.

As I told you, there are hundreds of types of carnivorous plants, and they employ special techniques to ensnare their insect prey—they snap, trap, stick, and suck.

The Venus flytrap uses a snap trap. I'm not sure why they are called flytraps. Flies only make up about 5% of their diet. No, they probably should be called spidertraps[14] or anttraps. No matter. The mechanism works the same for all.

Many insects are drawn to Venus's reddish, mouthlike leaves by the smell of nectar. Mr. Ant isn't looking for nectar; he's just in the wrong place at the wrong time. Each side of Venus's trap has three or four thin sensor hairs. Tripping these trigger hairs is what gets you in trouble. If Mr. Ant knew what we know about this clever trap, he could avoid a terrible death. It's unlikely that he's as well-informed, however.

Stepping onto the leaves of the trap, Mr. Ant triggers the first hair. This creates an electrical signal that travels along the surface of the trap, much like the electrical signal that travels through an animal's nervous system. Nothing happens. Yet. The energy of that first signal is stored, and Venus's internal clock is now ticking. What Mr. Ant does in the next 20 seconds will determine his fate. Unfortunately, he chooses to blunder along across the trap farther. In doing so, he sets off a second trigger. This touch also generates an electric signal. Together, the energy from both signals passes a response threshold. The trap only needs one-tenth of a second to respond. It snaps shut.

Things look bleak for Mr. Ant, but all is not lost. If only he keeps still. If no further triggers are tripped, the trap will deactivate, and Mr. Ant could scoot free[15] in about 12 hours. The

---

[14] I know. Spiders aren't insects. But the Venus flytrap does not discriminate.
[15] This is how the trap avoids snapping shut on raindrops or other false alarms.

likelihood that he, or any insect, could keep cool for 12 hours is not very high. Once Mr. Ant trips a third trigger, the lobes of the trap clamp down tighter, creating an airtight chamber. Mr. Ant begins to struggle, which trips more triggers. By the time he trips five triggers, he has doomed himself. The airtight chamber now floods with a digestive enzyme. Trapped and sealed off from oxygen, Mr. Ant will, sadly, asphyxiate and die.

The fluid in the trap is acidic and contains an enzyme that begins to break down Mr. Ant's body. It will take Venus several days to digest. The nutrients are absorbed by her leaves. After a week or so—after every bit of delicious ant essence has been sucked dry and he is no more than an empty husk—Venus will open the trap. Mr. Ant's lifeless remains will fall out or blow away. Venus will use this trap again a few more times before she replaces it with a new trap.

When I go check on my own Venus, I see that two of the little traps are closed. This, I think, is a good sign. I also notice that there is a pillbug (*Armadillidium vulgare*) perched precariously along the edge of my purple pitcher plant (*Sarracenia purpurea*). I feel a tiny bit of dread for the pillbug. I know what is about to happen.

THE PITCHER'S PLOY

Pitcher plants are like the couch potatoes of the carnivorous plant world. They literally *do* nothing. Their trap is a bucket made from a highly modified leaf. The curled leaf creates a deep, fluid-filled cavity. Foraging insects are attracted to the cavity by its smell and colors. The rim of the trap is slippery. It's very difficult, say, for a fly to walk along the edge without losing its footing. Once it falls in, its troubles begin. It lands in a puddle at the bottom of the trap.

Could that be rainwater? Not exactly. That liquid is a digestive fluid full of enzymes. It's aided by a type of bacteria that can help break substances down. It doesn't take long for the fly to figure out that this sticky, viscous fluid is not great for its delicate wings. Instinct will tell it to crawl out of there as soon as possible. Unfortunately, the sides of the trap are slippery too. In some cases, the walls are grooved. This greatly hinders the fly's ability to climb out.

Covered in goo and desperately dog paddling, the fly will eventually become exhausted and drown. Then the pitcher will take its time to digest it. Pitcher traps are physically the largest of all the carnivorous plant trap mechanisms; some can hold more than 1 liter of water and catch small animals such as frogs or rats.[16] I must turn away after seeing the pillbug fall into my pitcher plant.

I am too distracted to watch, to be honest. My attention has been diverted by my sundew (*Drosera filiformis*), commonly known as a thread-leaved sundew. Sundews stick. Like flypaper. That's their mechanism. The thread-leaved sundew looks a lot like its name—a collection of long, skinny leaves studded with short, red, hair-like tentacles. The tentacles ooze a sticky liquid made up of rainwater and complex sugars. These glistening droplets of "dew" are quite attractive to insects. My plant already seems to have intercepted a cloud of passing gnats. Those that attempted a landing quickly found their little feet all gummed up by the sticky

---

[16] And because our planet is such a strange place, it might not surprise you to learn that some pitcher plants thrive not on dead animals but on animal poop. *Nepenthes lowii* on the island of Borneo in Southeast Asia doubles as a shrew's loo. Its leaves are perfectly contoured to the mountain tree shrew's (*Tupaia montana*) backside. Shrews feed on the nectar coating the undersides of pitcher plant leaves. Then, quite conveniently, they poop into the pitcher. Nothing could please the pitcher more, as shrew poo is rich in nitrogen, a substance Borneo's soil sorely lacks.

droplets. Special glands at the base of the leaves then began releasing digestive enzymes onto their new guests.

A STICKY SITUATION

If you owned a Cape sundew (*Drosera capensis*), you would see something else start to happen—thigmotropism. That's the turning or bending of a plant in response to a touch stimulus. We don't expect plants to do this, of course, so when you see it happening, it's a bit unnerving. As an insect struggles to unstick itself, the cape sundew will curl its leaves around the insect. This creates more points of contact between it and the insect, and digestion takes place at a more rapid rate. My sundew has apparently been working hard all morning, as it is littered with gnat corpses.

Now if only I had a bladderwort (genus Utricularia) to complete my collection of sinister plants. Bladderworts are rootless, freshwater plants, and they claim the title of fastest carnivorous plants on the planet. They suck. Small, hollow, meat-eating bladders form along the stem of the plant. Each bladder comes equipped with a flexible valve that is lined with trigger hairs. When the valve is sealed, the plant pumps water out of the bladders, which lowers the pressure inside and sets the trap.

When a water flea or some other small creature swims past and trips the trigger hairs, the valve blows open. The low pressure inside sucks water and the animal in with an acceleration more than 600 times the force of gravity. Then the valve snaps shut behind it. The bladder now floods with enzymes, which help it digest its prey. Within 30 minutes, the trap is reset by passing water to the exterior and creating a new vacuum.

It's a lot to take in. Meat-eating plants. Animals you never knew existed. People eating things like cockroaches. On purpose. But there's so much more. Get this: When we're not busy eating them, insects are cleaning up our messes. They play a huge role as decomposers and recyclers. They are Nature's cleanup crew—feeding on decaying plant and animal matter, breaking down dead things into the building blocks of new life. Now that you've probably finished off your healthy cricket snack, you're ready to learn all about some of this unappetizing stuff in the next chapter.

4.

CLEANUP ON AISLE FOUR

THIS IS A chapter about death. And decomposition. General rot, putrefaction, and bad smells. It's a chapter about maggots and blow flies and dung beetles. I just want to let you know up front in case you don't feel up to reading it. If you think you can stomach it, though, I ask you to press on because the information in this chapter is extremely important to how life on Earth works.

The first thing that you need to know is that all living things die. Eventually. After they die, they decompose. We have a piece of family lore that relates to this very concept. My teenage sons sometimes regale their friends with it. At the time, Liam was five years old and Devin was three.

Picture a beautiful fall Sunday afternoon. Liam asked if we could go for a walk in the wilderness. We had recently moved away from Capitol Hill and into a house that faces Rock Creek Park—our wilderness. We were still getting to know our new neighborhood. Liam and I ambled up the street and went into the woods on the skinny, paved trail.

Before long, we saw deer tracks in the mud near the path. On a whim, we decided to abandon the paved trail and follow these tracks. We started to scale the hill but didn't get five paces before

Liam abruptly stopped. He tightened his grip on my hand and announced, "Something smells REALLY bad." I looked at him, and his eyes were wide. I smelled what he was smelling. "It smells like something *died*," he said indignantly.

I scanned the area and that's when I saw it: a buck in the advanced stages of decomposition on its side against a log—not more than 10 paces from us. I knew I had maybe eight seconds before Liam's eyes found it too, but I felt myself needing more time than that to know what to do. I spun him around to face me so he couldn't see it.

"Hang on a second, Liam," I said.

"What IS it?" he asked, already knowing that I discovered something.

"That smell you smell is a deer, honey. He died." Liam made a little whimpering sound. I picked him up, keeping him turned away from the carcass.

"Can I see?" he whispered. I weighed that in my head. I didn't know what was right to do. I stayed still for a second, holding him, mulling this over while the smell of death surrounded us. I didn't want to traumatize him.

"It's kind of gross," I told him gently, hoping to dissuade any interest. "He's almost all bones now."

"Okay," he said, bracing himself for it. "I'm ready."

I turned around slowly and took a step forward. "See?" I said. From the safety of my arms, Liam studied the body carefully. "Those are his ribs," I said, pointing to the most prominent part jutting skyward like sticky, splayed fingers.

"Why are they black?" he asked.

"Some of that is rotting skin," I told him, "which is what you are smelling. He's been dead for a while; most of him has rotted away, but some of him is still left."

"Why is he rotting?"

"All living things do when they die. See all the brown leaves on the ground? They are rotting too. And when people die . . ."

"They rot."

"Yes."

We stood there studying the deer for a while. Then I told Liam we needed to get back. It was getting darker and colder. I carried him down the hill and for a stretch along the trail.

"How old was he, Mommy?"

"I don't think he was that old. His antlers were small."

"Maybe a snake bit him, and he didn't know it, and he died from the poison."

We tossed out theories as we walked back. When we got to the house, we were both pretty cold. I put Liam on the couch and wrapped him in a blanket to warm up. I wondered what he was thinking. Our cat, Boone, saw the blanket and figured that was an invitation to snuggle up, so he jumped on Liam's chest and settled in.

Liam seemed so small on the couch under that blanket. He looked at me with his beautiful, almond eyes and asked, "Can he smell it, Mommy?"

"Smell what?"

"Smell the death on us."

"No, honey. Boone can smell that we were outdoors, but we don't smell like death. We didn't touch the deer, and we weren't that close to him."

He considered this for a minute.

"Are his eyeballs gone?"

"Yeah, I think they probably are."

"Can you see his brain now through his eye holes?"

"Well, his brain is probably gone now too."

"How long will it take for him to finish rotting?"

"I'm not really sure. He's been out there awhile, I'd guess, but he's got a little ways to go."

"Can we go back there? When he's done? To see his bones?"

I wasn't sure how to answer that one.

Two months passed. The dead deer in the wilderness was not forgotten. Liam asked about him often, wondering what his current condition was and how soon before we could go back to see his bones.

I decided one morning to investigate. I climbed the hill and searched for the deer. I found him all akimbo. His head was detached, and someone had sawed off his antlers. I was incensed. Who would have done such a thing? Did they not realize this was OUR corpse?

I stomped home. Over breakfast—because, of course, this is the *perfect* topic to discuss over breakfast—I told the boys what I discovered.

"Go get him, Mommy," Devin said.

"Yeah," Liam said. "We should go get him and bring him home to keep him safe. He doesn't like all his pieces being spread out."

"Get him, Mommy," Devin said again.

It hadn't really occurred to me to actually *retrieve* the bones. I thought maybe we'd just look at them one more time. But once I thought about it, I was warming up to the idea. It seemed like a really good idea, in fact. Why not? The bones were just sitting there anyway. The boys might find them interesting to look at. Maybe they would bring them to school when they were older. I realized I was feeling some sort of ownership of the bones, a protectiveness toward them. Shouldn't the deer pieces be kept together? Isn't there a spiritual feng shui about that? This was, perhaps, a moral obligation. I couldn't ignore the bones. It would be wrong. I NEEDED TO GET THE BONES.

GET THE BONES

Over bowls of Lucky Charms, we started plotting out our mission to retrieve the bones. Chuck was very quiet at his end of the table with a strange expression on his face. What was that expression? I wasn't quite sure.

After breakfast, he sidled up to me.

"You're not serious about the deer, are you?" he asked.

"Of course I'm serious about the deer. If we don't get him today, someone else might. Somebody already stole his antlers. Honestly, I don't know what is wrong with people. Why would you just take his antlers and not his whole head?"

"Why indeed?" Chuck murmured. That funny look crossed his face again.

Later that afternoon, Chuck put Devin down for his nap. Liam and I prepared to go. We made our way to the wilderness armed with a pair of Day-Glo dishwashing gloves, a black trash bag, and a Macy's shopping bag with handles. It took only a few minutes to find the body, just as I had left it this morning. Upon examination, I realized the skull wasn't quite as clean as a person might want it to be. It still had some stuff on it. Fur and rot, I guess. Hmmm . . . we would have to get that off somehow.

Donning the gloves, I opened the trash bag and grabbed the skull from the outside of the bag, dragging three vertebrae along with it. I gave it a hard shake to dislodge the maggots. We found one of the lower jaws in the leaves and added it to the bag. The hooves and legs appeared to be missing, and we couldn't find the other lower jaw. The rib cage was a few feet away, still a stinking mess. We discovered a lot of activity in the chest cavity; maggots and other critters industriously wove in and out of the rib cage. Interesting. Gross. Liam and I discussed.

At last, we felt ready to move on. I put the garbage bag in the bag with handles, and we found the path again. On my left side, I was holding hands with Liam. On my right side, I was toting a Macy's bag of foulness. This was normal, right?

We got back to the house and took the bag straight to the backyard. I guessed that Chuck wouldn't take kindly to me bringing it through the house. I left the head unattended for a moment, and we ducked inside. Now the hard part: Breaking the news to Chuck that the skull wasn't quite ready for display just yet.

Chuck was standing in the living room, waiting for us with arms crossed. "Well?" he said.

"Oh yeah," Liam said. "We got it. We got old Mr. Deer Head. He's in the backyard, Daddy. But we're gonna have to cook his head, Mom says. Right, Mom?"

Yup. We were gonna have to boil it. So much for breaking the news gently.

"Cook?" he managed to ask.

"Well . . . uh . . . yeah, I reckon so. Just to . . . clean him up a little," I answered.

"Daddy, he's got fur and ickies ALL OVER his head, so we've got to give him a bath. Right, Mom?"

"You're NOT cooking it in the house," Chuck said.

"NOT in the house," Liam and I agreed simultaneously.

It was sort of a strange problem to have, but after some thought, it was agreed that a Bunsen burner kind of thing could be rigged up in the backyard, and Mr. Deer Head could be stuffed into the big stew pot that we never used. While Chuck readied the flame, I tried stuffing Mr. Head in the pot. No go. There was too much connective tissue between the head and vertebrae. The vertebrae made the head too long, and it jutted out the top of the pot.

Hmmmm. Something had to give. I took the vertebrae in my gloved hand and puuuusssssshhhhhhhed it until I heard a click in the neck. Mr. Deer Head slid easily into the pot.

Feeling slightly queasy, I filled the pot with hot water and put it on the flame. Liam and I peered in. Suddenly, two sizeable beetles swam to the surface. I nearly jumped out of my skin—did not expect *that*. I guess I naïvely assumed that our piece of dead deer was bug-free after I shook all the maggots off. To think I was carrying that down the street in a Macy's bag with *beetles . . . eeeew*.

Without looking in the pot or making eye contact with me, Chuck suddenly declared that he needed to run an errand. He disappeared, leaving me and Liam to the task at hand.

"Smells *bad*, Mom," Liam observed calmly. I looked over at him. He took all this in rather well. As if all moms cook DEER HEADS in their backyards on the weekends. Silently, I prayed that I wasn't turning him into a serial killer.

What now? Mr. Head needed to stew a bit. Devin was still napping, so we went back inside to do regular things. Every few minutes, I went back out to see how it was coming. I did laundry, put fresh sheets on the boys' beds, checked the deer head, did the dishes, checked the deer head, vacuumed, checked the deer head. The pot was at full boil now. I saw my neighbor next door was out working on his yard. Surely he wouldn't come over, would he, and ask what I had going in the pot? What on Earth would I say? *We're just cooking up a little bit of DEER HEAD, sir. It's an old family recipe to get the sinew off.* But no, he seemed focused on his flowerbeds. I was probably safe.

Devin woke up from his nap, which is why I wasn't WATCH-ING when the pot BOILED OVER, sending deer bits all over our patio. It didn't take me long to discover it, though, because the SMELL overpowered us inside the house!

I dashed out to extinguish the flame, and the water settled down. The water was murky; it was hard to see if my 20-minute boil did any good. Chuck returned from his errand, looking at me warily. I confessed the boil-over and asked if he couldn't maybe take the kids for a walk while I dealt with the mess. Both boys protested, but Chuck was only too happy to leave again.

Which meant I was alone with The Head. With gloved hands, I reached in the hot pot and pulled it up. I felt my strength fail when I saw that it was still covered with fur and bits. I had hoped that all that stuff would just—fall off, I guess.

No dice. It was all still there. Which meant I was going to have to, you know, remove it. By force. Nothing beats the smell of dead, wet deer pelt. I felt my stomach lurch. It was time for some tough love: *You might be up to your elbows in steaming deer bits,* I told myself, *but what are your options? The situation is clear, and there's no turning back. You can't put it back in the woods. Soldier on.*

With renewed determination, I grabbed a piece of the steaming, waterlogged fur and peeeeeeeeeeeeeeeled it away from the skull. Oddly enough, it peeled fairly easily. In just a matter of seconds, I was left holding a soggy clump, which I daintily deposited in a trash bag.

The skull was mostly clear now. The vertebrae, however, were another matter. They weren't covered in fur but fused together with a dark brown substance. I didn't get the sense that was going to come off any time soon, no matter how many boilings I subjected it to.

No matter. I had gotten this far. I transferred all the bones to a bucket and started hosing them off. Then I attacked them with a scrub brush, making a mental note to throw this brush away when I was done and to *not* return it to the kitchen sink. In my final act as a junior taxidermist, I dunked everything into a bucket of hydrogen peroxide and water.

Then I waited. In theory, the hydrogen peroxide would bleach the bones without harming them. It would also sterilize them. I would let them soak overnight, then dry them in the sun.

I sat down, stared at the bucket, and thought to myself that breakfast was only six hours ago. Six hours ago, I had never boiled a deer head in my backyard before. Six hours ago, I had very limited experience, in fact, with touching dead things at all. Now I could probably be considered something of a pro.

It was a good idea. Hadn't I thought that back at breakfast over the Lucky Charms? Didn't I have some noble ambition about keeping the deer whole and having something interesting for the boys?

But that was hours and a whole lot of funk ago. I stared at the bucket. Maybe tomorrow all the stray bits of icky would have magically fallen off, and the bones would look clean and beautiful.

As it happened, Ol' Mr. Deer Head did, in fact, clean up rather nicely. But the experience was slightly traumatizing, so I waited a really, really long time before revisiting the site. By then, the bones were picked clean, and the insect world had moved on. I collected what I could to reunite Mr. Deer with the rest of his bones. The basket of bones sits in a place of honor in our home today—a constant reminder to the "weird things my mom does."

DECONSTRUCTING DECOMPOSERS

Things die. You don't normally find yourself in such close proximity to a juicy carcass like I did. But that's not to say that things aren't dropping dead all the time. Animals, yes, but plants too. And what happens to those things when they die? Nature needs to clean up after itself.

A good number of insects play the role of decomposer. You know how all this works, right? In a healthy ecosystem, you have producers, consumers, and decomposers. Producers make their own food. Think plants. Plants use light energy from the sun, carbon dioxide from the air, and water and nutrients from the soil to make food for themselves. Then consumers, who can't make their own food, come along and eat the plants. Not all consumers eat plants. Some eat other consumers. Some eat both. Decomposers have no interest in producers or consumers. Unless, of course, they happen to be dead. Decomposers accelerate the process of decomposition by physically and chemically dismantling organic matter. By breaking things down, valuable substances—like water, carbon dioxide, nitrogen, phosphorus, and calcium—can be used again.

Within the realm of decomposers are detritivores—animals that *eat* dead organic material.[1] Quite a few detritivores happen to be insects, which is why I mention it. The first group of detritivores we call xylophages. You're thinking: *Who is this "we" business? I don't know what a xylophage is.* Entomologists call them that. Now that we know this word, we can impress our friends with it. Xylophages primarily eat wood.

When a tree falls in the forest, I have no idea if anyone hears it, but I do know that some shredders are on their way. There are many soil- and wood-inhabiting insects that shred leaves or tunnel in dead wood. Beetles and pillbugs gnaw things into smaller, softer pieces. Some of this is eaten, and the rest is left behind for even smaller decomposers.

Nature isn't wasteful, you see. Decomposers ensure that no nutrients are lost. All that insect action helps plant matter decay quickly. Over time, decay creates humus,[2] a type of rich soil. Humus is an incubator for fungi, bacteria, and other microorganisms.[3] These, in turn, work through the soil, releasing the substances from whatever's rotting to be used as resources for living plants. Humus makes soil a little clumpy, which creates easier pathways for air and water to move through and allows oxygen to reach the roots of plants.

You can start to see why decomposers are critical in the flow of energy through an ecosystem. By returning nutrients to the soil, decomposers help new producers grow. Producers provide the nutrients that are passed to every member of the food chain.

---

[1] The difference here is that other decomposers directly absorb nutrients through external and biological processes while detritivores must ingest them.

[2] Not be confused with hummus, which is a thick paste made from ground chickpeas and is delicious with pita bread

[3] You can remember them as FBI. No, not *that* FBI. Fungi, bacteria, and invertebrates.

That's why trees getting chopped down and sent to the lumber mill on a large scale is bad for the ecosystem. The decomposers cannot do their job, and the forest loses the valuable nutrients that were in the trees.

TREE TASTING

Xylophages don't always look for dead leaves or wait for a tree to fall. Living trees can be perfect targets for some like the horntail. Unlike carpenter bees, horntails[4] (family Siricidae) actually do *eat* the wood in which they nest. It starts with the mother horntail. I'll wait here a second while you look her up. It helps to see her so that you can be appropriately afraid of her. She looks like evil incarnate. Many a coniferous tree trembles in her wake.

A lot of female insects use a special organ called an ovipositor to lay their eggs.[5] Well Madame Horntail does some next-level stuff with the way she goes about it. She seeks out a weakened or dying tree and *drills* her ovipositor straight into the wood nearly 2 centimeters (0.75 inches) deep. There she lays up to seven eggs, and while she's at it, she squirts in a load of fungus she keeps tucked away in her abdomen for such occasions. The fungus gets to work on the wood immediately, predigesting it for the arriving larvae, which hatch a few weeks later. The larvae are thrilled to find this soft, spongy wood to feast on. As they molt and grow, they tunnel deeper and deeper into the tree, following the trail left behind by the fungus.

Horntail larvae can spend a year or more chewing their way through a tree before they surface and transform into adults.

---

[4] Horntails are also called wood wasps.
[5] If you don't know about this, no worries. We'll get to that in Chapter Nine.

One female horntail can lay up to 200 eggs—you can imagine the trouble a poor tree might find itself in. Wood-boring insects can wreak quite a bit of havoc, as we'll see in the next chapter (Chapter Six: How Insects Suck), but for now, it's enough to know that you don't want to be the lone pine tree at a party of horntails.

You might be wondering why any living thing would want to eat wood. It can't have that much nutritional value, can it? It depends on who you ask. You and I won't get much out of gnawing on twigs, but many insects are built specifically for this diet.

You may not have known the word *xylophage*, but I'm guessing you know the word *termite*. Termites are xylophages. Before you get all judgy, let me just say that termites suffer from an image problem. Yes, yes, yes, they do *on occasion* eat people's homes,[6] but termites are master builders and incredible creatures. You should know that most termites are not pests. Of the 2,800 termite species in the world, only 28 have an interest in eating your house. Other species are known to aerate the soil with tunnel systems, to clear dead wood, and to boost plant growth.

Termites are hard workers. The structures they build—you can't believe it. The largest on record was two stories high. Did I mention they don't sleep? Never. Not ever. They build their colonies 24 hours a day, every day, until they die. And speaking of death, termite queens have some of the longest life spans in the insect world—up to 50 years. Some termite species' queens can lay 15–25 eggs per minute. That's more than 40,000 per day. Do the math!

---

[6] To the tune of $5 billion annually, not that we're counting or anything

There's so much that can be said about the termite. But you were wondering how it is that an insect can eat wood.

Wood isn't an overly nutritious thing. It's mostly made up of a complex sugar called cellulose. Cellulose is tough to digest, so most xylophagous insects need help doing so. Termites happen to have magic little bellies that are full of bacteria and protists. Protists are neither animal nor plant nor fungi; they are just these weird little organisms that swim around using taillike appendages called flagella. As near as we can tell, protists help termites break down wood by fermenting it. It's sort of the way we turn grain into beer or milk into yogurt.

Not all xylophagous insects have this special gut. Many wood-boring beetles have special enzymes in their systems to break down the cellulose. That's what the deathwatch beetle relies on. If you think termites face a lot of fear and loathing, try being a deathwatch beetle. I mean, it's hardly fair, is it? To saddle something with a name like that? They don't stand a fighting chance of being liked. Yes, okay, the deathwatch beetle does *sometimes* infest the structural timbers of old buildings. And it's true that the deathwatch beetle does have one tiny, *potentially* annoying habit of tap-tap-tapping its head on wooden beams when it's looking for a mate. That tapping sound is hardly a harbinger of death, yet some people insist that what is pure biology on the part of the beetles, is actually basis for superstition. In medieval European folklore, the nocturnal deathwatch beetle was thought to be an indicator of the approaching death of a member of the household in which the tapping occurred. This is likely because people would hear the tapping while sitting up with a sick or dying person during the night and think it was the sound of the grim reaper tapping his scythe on the door to take away the soon-to-be-deceased.

It's the stuff of nightmares, really. Even our man Edgar Allan Poe, who wrote many creepy stories, based one of his tales on the deathwatch beetle—his 1843 short story "The Tell-Tale Heart." If you've never read it, it's a good one. An unnamed narrator confesses to the reader that he killed an old man and hid the pieces under the floorboards.[7] The killer is slowly driven mad by a tap-tap-tapping sound coming from the floorboards. He ascribes the sound to the dead man's still-beating heart. Literary scholars have suggested Poe used the deathwatch beetle as his muse. How can you love this beetle after that?[8] Honestly.

TO EAT THE DEAD

Now we have a good understanding on xylophages. Let's move on. There's an entirely different category of detritivores called necrophages. You'll like them even less than the deathwatch beetle, I fear. Guess what they eat? Dead animals. Before you get yourself all hot and bothered by this, let me remind you: Everything dies eventually, and you can't have this stuff just piling up. It has to go somewhere. Necrophages provide a valuable service.

For icky but important behavior, consider the American burying beetle (*Nicrophorus americanus*). These little insects can pick up the stink of something dead within minutes of it being dead. Once a burying beetle finds a fresh, steaming carcass, it waits for reinforcements. The beetles want the carcass to feed their young, but it takes no small amout of cooperation to get the body ready.

First, they slice off all of Dead Thing's fur or feathers, which

---

[7] What did you expect? It is Poe, after all.
[8] It's worse if you're a Sandra Bullock fan and have seen or read Alice Hoffman's *Practical Magic*. Bullock's character hears the deathwatch beetle in her home and tries to dig it out of the woodwork before her husband meets an untimely end.

they'll use to line a pit. Then they roll Dead Thing into a neat little ball while simultaneously digging underneath it. One of the females will use this opportunity to lay a slew of eggs. Dead Thing slowly sinks into the pit, and the crew covers it. The entire operation takes 12–18 hours. When the eggs hatch about four days later, they wriggle to the corpse and begin feeding. After eight or nine days, the larvae have matured, by which time the carcass is reduced to bones. Yes, it's icky, but you have to admire the efficiency of it all.

CRIME FIGHTERS

We rely on necrophagous insects in other ways too. Their life cycles are so predictable, we use them to fight crime. The first documented case of insect crime fighting came from the 13th-century Song Dynasty in China. Song Ci[9] was a physician, judge, and investigator. He happened to catch an interesting murder case back in 1235. The victim had been dispatched with a slash to the neck. Song Ci ordered all the villagers to fetch their sickles—a sharp tool used for cutting rice at harvest time—and to place them on the ground in front of him. Then he waited. Not long. Soon a small swarm of flies collected above one sickle in particular. They were attracted to traces of blood on the blade. Song Ci interrogated the owner of that blade and soon had a confession.

As an investigator, Song Ci must have seen a lot of action, and he clearly took a lot of notes about his cases. In 1247, he sat down to pen a little something called *The Washing Away of Wrongs*.[10] He wanted to create a guide for other investigators that was rooted in practical observation. Song Ci outlined best practices for handling

---

[9] Also known as Sung Tzu
[10] Also known as *Collected Cases of Injustice Rectified*

dead bodies and collecting evidence. He explained how to examine bodies under certain conditions—buried versus non-buried, for example—and what to look for at various stages of decomposition. He covers a lot of ground in this work, including the usual stuff: blunt force trauma, knife wounds, drownings, and strangulation. And a few less common forms of death: fright from devils or goblins, death at the hands of mad dogs or tigers, cauterization, bambooing, trampling by horses, or being poisoned by tortoise flesh. Like I said, Song Ci must've seen a thing or two.

His book became a field manual for investigators working crime scenes. It was used in China for centuries and was translated into many languages, including English. You can still find copies of it today.[11]

The sickle murder was one of the world's first documented forensic entomology cases, but it would hardly be the last. Building off this important work, our understanding of how insects interact with the dead and dying began to deepen and evolve.

A LESSON IN ROTTEN MEAT

Let's leave 13th-century China and speed ahead to Italy circa 1668. It was at this time and in this place that Francesco Redi made a very important discovery. Redi was an Italian naturalist and physician who's best remembered for challenging Aristotle's idea of abiogenesis—that's the notion that living organisms could arise from nonliving matter. Redi's particular interest was in rotting meat. He, like many others of his time, had observed that where there was rotting meat, there were often maggots. It was generally assumed that rotting meat *caused* maggots. That idea didn't sit right

---

[11] I recommend Brian McKnight's 1981 translation.

with Redi, so he set out to disprove this concept of spontaneous generation. He did so with a clever experiment we like to call "meat in a jar." Sounds icky. And it is.

Redi gathered up some common household items—meat and jars. He filled two sets of four jars with different types of meat and fish. Half the jars were left open; the other half were sealed with paper and string. The fish and meat in both groups rotted, but maggots only appeared in the open jars. Redi repeated the experiment with jars covered in fine gauze. Here, the meat rotted, but no maggots appeared in the jars.[12] Redi realized that the maggots weren't coming *from* the meat. They were coming from flies that were *attracted* to the meat and were laying eggs there. His conclusion: "omne vivum ex vivo" or "all life comes from life." You might think this is pretty basic stuff, but at the time, this was breaking-news science.

MYSTERY OF THE MUMMIFIED BABY

It was Dr. Louis François Etienne Bergeret, a French physician, who took another giant leap forward for the world's would-be forensic entomologists and murder squad detectives when he was handed what has been dubbed the Parisian mummified baby case. The year was 1855. A couple remodeling their Paris home made a shocking discovery. Hidden inside the brickwork of their fireplace was the mummified remains of an infant. Naturally, suspicion immediately fell on the couple. It was their home, after all. The couple professed their innocence. *We just got here*, they said. *We don't even have any kids.*

---

[12] The experiment was also an important one for the scientific method because it used a control group to test a hypothesis.

Not anymore, the detectives replied.

The infant did display some disturbing signs of insect infestation. By the 1800s, scientists had a fairly good understanding that insects would inhabit decomposing bodies. What they were trying to suss out, though, was the idea of succession: When one group of insects arrives and feeds, it simultaneously paves the way for another. But which insects came *first*? And how long did they stay? For detectives, could this information be useful in trying to determine a person's time of death?

Enter Dr. Bergeret. The task of performing an autopsy on the mummified remains fell to him. Dr. Bergeret knew a little something about insect life cycles. Based on what he found on the body, he concluded that it had been placed behind the mantel years earlier in 1848. This revelation certainly got the new homeowners off the hook. Things didn't look so good for the couple who had been living there back in '48, though. They were charged and subsequently convicted of murder. This case marked the first time that insects were used to estimate a person's time of death.[13]

In the 1890s, French veterinarian Jean Pierre Mégnin continued to pursue the idea of succession. Like Dr. Bergeret, he was asked to aid in a murder investigation. Another infant. What would a vet know about murdered babies? Nothing. But Mégnin was a wizard acarologist. (I don't expect you to know what that is. It's someone who studies mites.) The body of the infant was covered in them, you see. The mites were in multiple stages of development.

In an experiment, Mégnin counted the numbers of dead mites

---

[13] This is called the post-mortem interval (PMI). It is the time that has elapsed since an individual's death.

and live mites that developed every 15 days and compared this with his initial count on the infant. Then he was able to estimate how long the infant had been dead. Mégnin wrote many articles and two important books[14] on forensic entomology. Based on his studies, he asserted that you could map out which insects might come to a dead body and how long they would stay. He posited that exposed corpses were subject to eight successional waves of insects. Buried corpses—trickier for insects to gain access to—were subject to only two waves.

Other scholars worked on this concept over the years, but it wasn't until the mid-1960s that a graduate student in North Carolina named Jerry Payne began laying the groundwork for the current theory of succession. His idea—as observed through a series of dead pigs—was that as each wave of insects feeds on a body, the body changes. These changes make the body more attractive to the next wave of insects. In this way, you can observe a predictable pattern with different groups occupying the body at different times. Insects from the orders Diptera (flies) and Coleoptera (beetles) will be most plentiful. Which insects appear and when is dependent largely on their needs for food and shelter, but the external conditions of the body—weather, temperature, location—can have an impact as well.

RUXTON'S FOLLY

Some very famous criminal cases have relied on the predictability of insects. An early example of such a case is the riveting double

---

[14] My high school French is not solid enough to help me decipher these, but *La Faune des Tombeaux* and *La Faune des Cadavres* are considered to be among the most important forensic entomology books in history.

Jigsaw Murders from 1935. It centered on the case of Dr. Bukht-yar Chompa Rustomji Ratanji Hakim, an Indian-born, British physician better known as Buck Ruxton. Dr. Ruxton was the jealous sort. His common-law wife, Isabella Kerr, was vivacious and social but *not* unfaithful as Dr. Ruxton imagined. Despite this, he strangled Isabella with his bare hands one night in a fit of rage. Unfortunately, the murder was witnessed by their maid, Mary Jane Rogerson. So he murdered her too.

Dr. Ruxton took great pains to conceal what he had done. To obscure his victims' identities, he cut the bodies into pieces and wrapped them in old clothing, bedsheets, and newspapers. He drove 160 kilometers (100 miles) north of Lancaster and dumped the pack-aged remains in a stream near the town of Moffat in Scotland.

A few weeks later, the first of the 70 parcels were discovered by poor Miss Susan Haines Johnson, visiting from Edinburgh. She looked out over the bridge, admiring the view, only to spot a par-cel with a human arm jutting out from it. Pandemonium ensued as detectives searched the area for more remains. Miss Johnson was probably never the same.

Doctors were called in to help reassemble the bodies. It was clear to them that the killer was someone who understood anat-omy and either had surgical skills or knew their way around a butcher's block. This narrowed the suspect pool considerably.

Dr. Ruxton thought himself to be quite clever. Yet, like a lot of criminals, he was not the sharpest tool in the shed. The news-papers he had chosen to use came from a special edition of the *Sunday Graphic* that sold only in the Lancaster area, which brought detectives to his door. Some of the clothing he had used was eas-ily traced back to his family members. Which, frankly, was just sloppy on his part. In addition, his home was chock-full of blood stains and other evidence. As the noose tightened, Dr. Ruxton foolishly gave multiple, conflicting stories to explain the absence of his wife and maid.

If that weren't enough—and surely it should have been—detectives used some fledgling forensic techniques to confirm the victims' identities[15] and establish their times of death. Maggot specimens from the parcels were sent to Dr. A. G. Mearns at the University of Edinburgh. Dr. Mearns, something of an expert on insects, was able to determine the date on which the body parts had been dumped in the river based on the presence of third instar[16] larvae of *Calliphora vicina* (bluebottle larvae).

Although Dr. Ruxton never confessed to his crimes, the weight of the physical evidence was enough to easily convict him. Dr. Ruxton was hanged for his crimes. The prosecution of his case would prove to be one of the United Kingdom's most publicized legal cases of the 1930s.[17] The maggot evidence, in particular, set legal precedence.

ALICE: A CASE STUDY

So how *does* all this work, you're wondering. Me too. I decided to find out by enrolling in an entomology course offered by the University of Alberta called Insect-Human Interactions. Led by a team of entomologists, the course was designed to cover all manner of intersections between people and insects. One full unit was devoted

---

[15] A forensic anthropologist superimposed a photograph of Isabella Kerr over the X-ray of one of the skulls that was found to prove that she was one of the victims.
[16] An instar is the phase between two periods of molting.
[17] It even inspired a sort of grim rhyme—*Red stains on the carpet, red stains on the knife / For Dr. Buck Ruxton had murdered his wife / The maid servant saw it and threatened to tell / So Dr. Buck Ruxton, he's killed her as well.*

to detritivores, and that's where I learned all about the five stages of human decomposition. My professors were rather clinical about the whole thing, so to make sure I was really, *really* understanding things, I also spent time reading briefs on countless forensic entomology cases and making friends at the American Board of Forensic Entomology and at Aftermath—"specialists in trauma cleaning and biohazard removal." Boy, do I have a lot to tell you! And you know by now that I'm going to tell you *everything*.

Here's my take: Let's say someone has just died. We'll call her Alice. Alice looks pretty lifeless, but there's actually a lot going on with the poor dear. As soon as blood circulation and respiration stop, Alice's body has no way of getting oxygen or removing wastes. Excess carbon dioxide causes membranes in her cells to rupture. Eeew. Yes. I know. Very messy. The membranes release enzymes that begin eating her cells from the inside out. Alice has entered the first stage of decomposition: autolysis, or self-digestion. Her body, at this stage, is considered fresh, but it won't stay fresh for long.

Over the next 36 hours—give or take—the body will pass through three mortises: algor mortis, rigor mortis, and livor mortis. Algor mortis just means that the body temperature will drop until it matches the temperature of her surroundings. During rigor mortis, the body will become stiff due to severe chemical changes. Livor mortis refers to the pooling of blood on the side of the body closest to the ground. If we were detectives, we might be looking for these stages to help us determine what happened to Alice and when. But even before that, something important happens.

As little as four minutes after her passing, Alice is met by her first visitors: flies. And the reason for this is because Alice stinks. I don't mean to suggest that Alice neglected her personal hygiene.

Far from it! In life, she always prided herself on her smart appearance and lovely scent. Gardenia, it was. That was her favorite. But now that she has gone to meet her maker, Alice no longer smells like gardenias, especially to flies.

No, she will now give off a distinctive, almost sickly sweet smell that the flies will be able to detect. This smell will change over the course of her decomposition, as it's made up of hundreds of volatile organic compounds. These compounds are produced by the actions of bacteria, which break down her tissues. You and I cannot smell Alice yet, but insects can smell her from more than 1.6 kilometers (1 mile) away.

The first wave of colonizers will likely be female blow flies, flesh flies, and house flies. They're here to feed on Alice's bodily fluids like blood, sweat, or tears and to lay eggs. I know, I know. She would simply be appalled. The flies first target body openings like Alice's eyes, nose, and mouth. These are easy access points. If she had any open wounds like bullet holes or tears in her skin from knife wounds, the flies would flock there too. During the first hour of Alice's inertness, these flies are going to lay a lot of eggs. A single blow fly might lay as many as 500. Each tiny egg is no longer than 2 millimeters (0.08 inches).

Within the first day of Alice dying, the eggs will pass through their first stage of development and hatch into maggots. This is the first instar. These maggots are about 5 millimeters (0.2 inches) long. They immediately move inside the body to begin to feed. Yes, on Alice. Sorry.

The outer layer of a maggot is called a cuticle. It's made from chitin,[18] which, as we know, cannot grow. As the maggots gluttonously

---

[18] We had this in Chapter One. Chitin is the same business that makes up an insect's exoskeleton.

indulge, they get a little too big for their britches and have to molt. If we were detectives and found Alice now, we'd see the molted casings and growing larvae.

Days two and three of Alice's life after death see more of the same. The maggots are hard at work. Eating and molting, eating and molting. By the time the maggots have reached their third instar, they are rather large: 17 millimeters (0.67 inches). Alice would be deeply disturbed to realize that she is no longer a cold corpse. There's so much activity on her body from maggots feeding, her core temperature has risen by as much as 10°C (50°F).

I'd like to tell you that maggots are the worst of what she has to endure, but you have to figure that this pulsating mass of hot maggotry is not going to go unnoticed. A second wave of insects soon arrives. These are predatory insects, and they are less interested in Alice herself as they are her maggots. Say hello to the rove beetles, carrion beetles, Hister beetles, and checkered beetles. They begin gobbling maggots with wild abandon. Meanwhile, cheese flies[19] and faniid flies arrive.

Alice's body is changing as she enters the second stage of decomposition: putrefaction, which is a decidedly more icky-sounding word. It's also known as the bloated stage. Alice would be so distressed to hear this. She hated feeling bloated.

Unfortunately, her body does bloat[20] due to gases produced by bacteria. Her organs are beginning to liquefy, and she's becoming

---

[19] I know you're going to ask me why they are called cheese flies. I flirted with the notion of just pretending I hadn't actually mentioned them. Maybe your eyes would skip over that part. But no, you're too astute for that. When they aren't chowing down on corpses, you'll find live cheese larvae in casu marzu, a traditional Sardinian sheep's-milk cheese. And yes, people eat it.

[20] The body can, in fact, double in size.

quite odoriferous.[21] There's stuff happening to Alice's beautiful, porcelain skin that I simply cannot describe here.[22]

About this time, any maggots that weren't gobbled up by beetles have finished molting. It's time for them to leave. They want to find a dry area to finish their transformation into adulthood. So, they bid adieu to Alice, crawl away from her body, and burrow into the soil. This is another place detectives might look after Alice's body has been removed.

In the soil, the maggots form pupae. When they emerge days later, they are adults. Five days from now, they will start reproducing. Not on Alice. She's busy. They'll have to find a new sticky corpse.

Alice has already entered the third stage of decomposition: black putrefaction or active decay. Seems odd to call it that when there's already been so much activity. Here, what's left of Alice starts to turn black, and the gases that were building up during the bloat escape. A large volume of liquified body parts drain out and seep into the surrounding soil. Other insects and mites feed on this material. The larvae of the cheese flies are the only remaining maggots. Without as many maggots to feast on, the beetles take a more active interest in Alice.

Oh dear. Alice has been here for more than 10 days now. She has reached stage four: butyric fermentation or advanced decay. We would have a hard time recognizing her if we came upon her now. Most of her flesh is gone. The butyric acid makes Alice smell a bit yeasty. Another wave of beetles is drawn to this smell, as are

---

[21] The smells are so potent at this stage of decomposition that they can linger long after a body has been removed.

[22] Skin slippage. It's just not something you discuss in polite company.

predatory mites. Then the beetles lay *their* eggs on what's left of Alice. Their larvae will hatch and feed on her too.

Hair, bones, and bits of cartilage are all that remain of Alice in stage five: dry decay. More than 50 days have passed, and most insects have abandoned her. Insects that feed on hair—like tineid moths—appear. Keratin-feeding beetles in the family Dermestidae[23] come too. These scavengers feed on proteins that few other organisms can digest, such as skin, hair, and feathers. If Alice was wearing her feather boa, those beetles would be all over it. Thankfully, she is not. It's still at home in her closet. Eventually, all that will be left of Alice will be her bones. And they, too, will one day disintegrate.

It's a bit sad, isn't it? But what has happened to Alice happens to us all, eventually. She no longer needed her body. Knowing Alice as we did, we might imagine that she could find peace with it. Her remains fed thousands of insects that will become food for other creatures or that will perform other services in our ecosystem. What's more, they tidied up what was becoming a ghastly mess. And Alice, she so appreciated tidiness.

What if we found Alice right away and whisked her off to a mortuary and had her quickly embalmed? What if we buried her in a lead coffin, six feet underground? That would stave off things for a little while but not forever.

What if Alice had been taken from us prematurely? If she was

---

[23] Dermestid beetles are so efficient at removing organic material from a body that colonies are often kept in museums to clean the flesh and hair or feathers from vertebrate skeletons for display or research collections. Dermestids consume everything but bone, resulting in an almost perfectly clean skeleton.

murdered? Then we would want justice! We would want a forensic entomologist[24] to collect the proper evidence.

It's tricky, this. While insects are predictable, you must remember that a number of things can affect their development. Insects are vulnerable to extreme heat and extreme cold because they cannot regulate their own body temperatures. If it becomes too cold, insects can't fly and won't lay eggs. Larvae won't grow. High temperatures can increase the rate in which some insects develop; the warmer it becomes, the faster they grow.

To estimate Alice's time of death, we would need to do several things. First, we would need to collect specimens of every insect we can find on her body as well as any insect in or on the surrounding soil. Some of these would be immediately preserved; others would be kept alive and reared to adulthood. This is because many maggots look alike. It is only after they reach adulthood that they have more distinguishing features and can be more accurately identified. A forensic entomologist needs data from the crime scene to make any determinations because so many variables affect decomposition and must be taken into account. Was the body found inside or outside? What were the weather conditions? Was it cold? Was it rainy? Was it a hot summer day with a scorching sun? Was the body submerged in water? Was the body buried?

---

[24] If we know the words *forensic entomology*, most of us are quick to think of dead bodies and crime. To be fair, forensic entomology encompasses three fields. Medicolegal entomology involves using insects to solve violent crimes or to investigate sudden or suspicious deaths. Urban entomology involves insects in human-made structures. Cases like these might revolve around insect infestations. Stored-products entomology is a field you might already be acquainted with based on our previous chapter together. These cases deal with insect infestations in stored food and aims to determine when and from where insects invaded the food.

THE WORK OF DR. BILL BASS

We owe much of our current understanding of how insects interact with dead bodies to Dr. Bill Bass. Dr. Bass is a forensic anthropologist at the University of Tennessee. In 1971, he created what would become the famous Anthropological Research Facility, also known as the Body Farm.

Dr. Bass was inspired to create such a thing while he was a professor at the University of Kansas in the 1960s. He had been approached by the Kansas Bureau of Investigation on a cattle rustling case. The KBI needed Dr. Bass to determine the time of death of a partially decomposed cow. At that time, what we knew about decomposition and insect succession left us with more questions than answers. Dr. Bass felt like he couldn't be of much help in this case, but it did spark in him an idea. He wondered how something like that could be studied.

In the late 1970s, Dr. Bass found himself being presented with another interesting puzzle. He was summoned to a grave in Franklin, Tennessee, to investigate what was believed to have been a fresh homicide buried on top of the grave of a Confederate soldier—Colonel William Shy. Killed in 1864 during the Battle of Nashville, Shy had been laid to rest in a cast-iron coffin. It appeared that someone had reopened Shy's grave and deposited a new body on top. Due to the fact that this body was relatively intact and still contained most of its flesh, Dr. Bass initially estimated that the body had been dead for less than a year. His estimate was off by more than 100 years.

The body was, in fact, Shy's. Grave robbers had come along, punctured the coffin, and tried to remove him while looking for artifacts they could sell. Because the coffin had been airtight and underground, decomposition had been largely slowed. It was Shy's clothing—perfectly stylish for the 1860s but less so for the 1970s—that tipped Dr. Bass off that he might be dealing with something much older.

Dr. Bass was mortified by his error, but it made him realize that the lack of research in this area could be detrimental to investigators. So with his university's support, he began his research using unclaimed bodies from the county morgue.[25] It was certainly unusual and unorthodox work. Dr. Bass has buried bodies in shallow graves. He has submerged them in water. He has wrapped them in plastic. He has locked them in car trunks. He has left them out in the sun. With each case, he thoroughly documented the decomposition of the body and recorded what lived on it and when. Over time, he created a body of work that investigators have come to rely on.

DIVING FOR EVIDENCE

As Dr. Bass discovered, some crime scenes are harder to work than others. One summer in 1989, scuba divers made a startling discovery in the Muskegon River in western Michigan. They found a submerged car, which contained the body of a woman. That alone made things challenging. It's not like blow flies have any diving skills.

The investigators on the case suspected foul play from the beginning. The injuries to the woman's head were not consistent with a car accident. Instead, she looked like she had been clubbed. They tracked the car to the woman's husband. The husband claimed that he and his wife had an argument in June, a few

---

[25] A surprising number of people have donated their bodies, after death, to his research.

weeks before her body had been discovered. He said she was upset and had driven off, and he had not seen her since. The explanation didn't sit well with detectives,[26] but it was his word against hers, and she wasn't talking.

They looked to the evidence. Submersion in water typically slows decomposition. The cold water had preserved the wife's body, therefore estimating her time of death was difficult. Difficult but not impossible. Detectives collected a number of aquatic insect pupal cocoons and larval cases from the car's windshield, fender, and door handles. Then they asked Dr. Richard Merritt, an aquatic insect entomologist from Michigan State University, to identify and date them. He identified caddis fly cases, chironomid beetle larvae (family Chironomidae), and black fly larvae and pupae of the genus Praimulium. It was the black fly larvae that helped solve the case.

In this region, black flies lay eggs in late spring and early summer. The eggs hatch in rivers and streambeds in the late fall or early winter of the next year. Each larva has a circle of hooklike structures on its backside. It blurps out a clump of silk onto the nearest rock (or submerged car door handle) and clamps those hooks down to keep itself from being washed downstream. Now its head is free to eat whatever small thing passes by. Here it stays and grows during the cold winter months.

In March or April, the larvae pupate, emerging from their cocoons in early to mid-May. Adult black flies feed on blood to gain enough energy to mate and lay eggs. The flies die a month or two later. Based on what he knew, Dr. Merritt determined

---

[26] The fact that the woman's husband had taken out an insurance policy on his wife's life and already pawned all her jewelry did not make him look any less guilty.

that the car must have gone into the water *nine months before* the discovery of the woman's body. She had not disappeared after an argument a few weeks prior. She had disappeared in the fall of 1988.

The prosecutor presented evidence that the husband had clubbed his wife to death and sent her car careening into the river to look like an accident. Dr. Merritt's testimony helped convince the jury of the timing of her death. The husband was swiftly convicted of second-degree murder and handed a long prison sentence.

INSECTS DON'T LIE

Have you ever wondered how the medical examiner on detective shows always seems to know if a body has been moved? I wondered about that too. Sometimes they can figure it out based on how blood has pooled in a body. Other times, they know it based on insect evidence.

M. Lee Goff has a lot to say about this. Goff is a forensic entomologist and one of the founding members of the American Board of Forensic Entomology. He described a case in Oahu, Hawaii, where the decomposing body of a woman was found in a sugar cane field. Most of the insects on her body came from the island, but there was one particular maggot—*Synthesiomyia nudiseta*—that's found only in urban areas. That made Goff suspicious. Further, while the urban maggots seemed to be about five days old, the other maggots seemed to be around three days old. He suspected the woman had been killed in the city, and her body had been moved to the sugar cane field.

As it turned out, insects don't lie. Once the victim was identified, detectives pieced together her story. She had been killed in an apartment in Honolulu. The murder had been unplanned—a drug deal gone wrong. The murderer didn't know what to do, so he kept her body in place for a few days before deciding to try to hide it in the field. In this case, the insects not only told us *when* but also *where* the woman had been killed.

That case was fairly direct, but not all cases are as cut-and-dried. Investigators ran into a puzzler in Danville, Virginia, in 2006. Jonathan Blackwell was last seen leaving work at the Goodyear plant on October 7, 2004. His family reported him missing four days later. Eight days after his disappearance, the charred remains of his car were discovered. The trail stopped there.

Then in December 2006, authorities received an anonymous tip that Blackwell had been murdered in 2004. The tipster instructed investigators to search for a shallow grave near a barn in North Carolina. They did so and discovered Blackwell's body. His skull had been fractured, and it was later determined that he had been bludgeoned to death with a rock. But there was a problem: If Blackwell had been murdered in 2004, why was his body covered in seven-day-old blow fly larvae?

The case was sent to Wes Watson, a professor of entomology at North Carolina State University. Watson deduced that the shallow grave was not actually Blackwell's first grave. He *had* been murdered in 2004. The killer covered the body in lime before burying it with the mistaken belief that the lime would hasten decay. Not only does lime prevent the growth of microorganisms and inhibit decomposition but burial also initially blocks insects from colonizing the body. That's because even a few inches of soil covering a body will prevent blow flies from laying their eggs. The depth of burial, the nature of the soil, and the temperature and moisture content of the soil all affect decay. When the killer dug the body up to rebury it elsewhere, there was still a lot of tissue in place for blow flies to colonize. There was a lot more to this case, but eventually the killer confessed and was convicted.

INSECTS IN ABSENTIA

It was, in fact, the *lack* of insects on a body that freed one innocent woman of conviction. But not, unfortunately, before she had served 16 years in prison for the crime she did not commit. Kirstin Blaise Lobato was only 18 when she was accused of murdering and mutilating Duran Bailey, a homeless person in Las Vegas, Nevada. The murder took place in July 2001—long before DNA evidence was widely collected and used as evidence in trials. Through a series of inept legal maneuvers, Lobato was convicted of the crime despite having a strong alibi. Part of the case hinged on Bailey's time of death. The prosecution made several estimates and ultimately settled on a wide window that left a potential gap in Lobato's alibi.

It seems incredible that she could have been convicted based on what we know about insect succession. Yet it took the testimony of forensic entomologist Gail Anderson to free Lobato.

Anderson is the associate director of the School of Criminology at Simon Fraser University in British Columbia, Canada. What struck her about the case was the missing maggots. By all rights, the crime scene should have been teeming with flies. The body was exposed outdoors, and the body was covered in wounds and therefore quite bloody. The crime scene was discovered at 10:15 p.m. the night of the murder. The prosecution argued that the killing could have taken place 1 to 12 hours before discovery. And yet. And yet, viewing crime scene photos many years after the fact, Anderson saw no sign of blow fly eggs, which would have most certainly been present if the body had lain unattended for up to 12 hours. As you and I know, blow flies are diurnal—they are active during the day, and they rest at night. Based on this, Anderson theorized the murder took place after sunset and the body was discovered soon after death—at a time when Lobato was at home with her family three hours away.

Anderson's testimony was corroborated by two other entomologists. Based on this testimony, a writ of habeas corpus—indicating that Lobato was falsely imprisoned—was finally issued, and she was exonerated. The wheels of justice turn slowly. And in this case, justice relied on maggots.

I'll just say this. Insects are so good at crime fighting, they just have to show up at a murder scene to make a conviction. Goff tells the story of a woman who was murdered in Texas in 1985. Among the evidence collected at the scene was one mangled grasshopper found in her clothing. The police brought several suspects in for questioning. During a search of the suspects, you'll never guess what was found in the pant cuff of one of the suspects: the left hind leg of a grasshopper.[27] Coincidence? I don't think so! The left hind leg was, in fact, the only part of the grasshopper missing from the one found on the woman's body. When the body and the leg were reunited, the fracture marks lined up perfectly. The suspect was convicted of murder.

In another notable case, the grasshopper that determined the

---

[27] For the record, if the hind leg of a grasshopper was found on my clothing, I would confess to anything just to have it removed.

outcome of a murder trial was found neither on the victims nor the killer—it was found on the killer's car. What made it so remarkable was that it was the *wrong* grasshopper.

Vincent Brothers, a former vice principal of an elementary school in Bakersfield, California, was accused of murdering his estranged wife, three children, and mother-in-law in 2003. Brothers proclaimed his innocence and backed it up with an alibi: He could not possibly have committed the murders because he was in Ohio visiting other family members at the time the murders took place. FBI agents assigned to the case had a different theory. They were convinced that he rented a car and drove from Ohio to California to murder his family. The defense claimed his rental car never left Ohio.

The feds set out to disprove Brothers's alibi by removing the air filter and radiator from the rental car and sending it to Dr. Lynn Kimsey. Dr. Kimsey was the director of the Bohart Museum of Entomology at the University of California, Davis. She was asked to identify any mashed insects she found and indicate where they were from. It was a strange request, but Dr. Kimsey set to work.

What she found condemned Brothers. *Xanthippus corallipes pantherinus* is a large grasshopper native to the Great Plains region on the eastern slope of the Rockies. *Neacoryphus rubicollis* and another specimen, genus *Piesma* (family Piesmatidae), are only found in the West—in Southern California, Arizona, and Utah. *Polistes aurifer* is a paper wasp found in Arizona, California, Oregon, Nevada, Washington, Idaho, and Utah. Nothing from Ohio. These insects were consistent with what might be found on two major routes to get to California from the East. There were no butterflies, such as painted ladies or sulphur butterflies, which indicated the car wasn't driven during the day but at night. Brothers's rental car, which he claimed never left Ohio, registered some 4,500 unaccounted-for miles.

Dr. Kimsey was one of 137 witnesses called to testify in the internationally publicized case. She spent five hours giving testimony on the witness stand. The jury found Brothers guilty of five counts of first-degree murder. He was sentenced to the death penalty.

LEAVING THEIR MARK

In yet another famous case, a killer was convicted after insects left their mark on him. An unfortunate series of events left a young woman dead in Thousand Oaks, California, in 1982. Her body was discovered under a large eucalyptus tree near a field of wild oats. She was found at night, and a team of investigators canvased the scene until the early morning hours.

Later that morning, one of the sergeants from the sheriff's office noticed a number of inflamed bites on his ankles, waist, and bum. He thought they might be chigger bites. As it happened, he was not alone. *Most* of the response team had been bitten. Chigger attacks in this part of California were not common, and there were no bites on the victim.

Tellingly, one of the suspects the team came across had similar red welts on his lower legs, waist, and bum. The suspect claimed to have been bitten by fleas at his sister's house. So began the battle of the insects: flea or chigger?

That's how Dr. James Webb of the Chigger Research Laboratory[28]

---

[28] Yes, apparently there is such a thing.

of California State University came to be involved in the case. Chiggers are the larvae of mites from the family Trombiculidae. As parasites, they latch on to a host for a single blood meal before dropping to the ground and tunneling into the soil to complete their life cycles.

Dr. Webb and his team searched the soil of the crime scene to look for chiggers. Chiggers have quite specific requirements for their transformation into adulthood—soil pH, temperatures, and relative humidity all must be just so for the chiggers to be content. What this means for investigators is that the presence of chiggers can be quite specific to a given location.

As you might imagine, Dr. Webb's team did find chiggers (*Eutrombicula belkini*) in the soil beneath where the body was found. There were no chiggers present at the suspect's sister's property. The suspect had indeed murdered the young woman and, in an attempt to conceal her body, he covered her with brush and had been bitten by the chiggers in the process. He was convicted and sentenced to life in prison without parole.

You and I could stay in Sherlock Holmes mode for a lot longer; there are many, many cases on the books that rely on the testimony of insects. But we need to press on because there's one last category of insect decomposers that we really must discuss. We've talked about insects that consume decaying plant matter (xylophages), and we've look into those that eat dead animals and people (necrophages). Now we have to talk about insects that eat poop (coprophages). We have to.

AN UNPLEASANT DIET?

You might be feeling a little sensitive about this because it really is as disgusting as it sounds. Before we talk about poop eaters, we first need to talk about poop. Why would any animal be interested in eating another animal's droppings?[29]

Do you know anything about how cows digest food? Maybe not. That's okay. Let me give you a quick primer. When a cow takes a big mouthful of grass, very little of it gets chewed before it's swallowed. A cow's stomach has four chambers. The largest is called the rumen; it's about the size of a large trash can and works like a food processor. After filling up on largely unchewed grass, the food gets passed back and forth between the first two chambers as bacteria stored there tries to break it down. This is formally called rumination, although most of us know it as chewing the cud. It can take a while. As ol' Bessie chews her cud, it slowly moves on to the other two chambers and then through her intestines.

You would think that with all this digestive activity, Bessie's dinner would be fully processed. But no. In fact, a substantial amount of her stomach contents is only ever semi-digested. There's enough nutritional value in Bessie's poop to create a veritable feast for insects.

Thousands of species of coprophagous insects have been identified, especially in the fly and beetle families. You may know of one already: the dung beetle. I used to think that the dung beetle was a singular thing, but there are actually about 8,000 species of dung beetles. They all have one thing in common—they eat poop—but they go about it in different ways.

PHILOSOPHY OF POO

Dung beetles, it would seem, have a whole philosophy of poo, and as such, we can break them down into three distinct groups:

---

[29] *Lots* of animals do this, you know. Bunnies. Cute little bunnies eat their own poop on a regular basis. Just saying.

rollers, tunnellers, and dwellers. You've probably seen photos or videos of the rollers (telecoprids). When they come across a steaming pile of poo, two beetles—a male and a female—shape some of it into a ball. The male pushes the ball away from the pile with his back feet, and the female rides on top of the ball, hanging on for dear life. The beetle must make haste, you see, in case they are intercepted, and their ball gets stolen.[30] The goal is to find a safe place to bury it. The balls they make are either used by the female to lay her eggs in (called a brood ball) or as food for the adults to eat. Storing the dung underground keeps it fresh. Mom usually hangs around the brood ball for a few months to clean the grubs after they hatch. When the larvae hatch, they have an instant food source and a fierce protector.

Tunnellers (endocoprids) find a steaming pile of poo and dive right in. Their goal is to tunnel straight through it and into the

---

[30] I said there were three groups. There are actually four if you count the kleptocoprids. Think of kleptomaniacs—they steal poo balls from the rollers. How low can you get, right?

soil. While Mom is setting up the family home down below, Dad goes up and down, carrying mouthfuls of poo to store as food. Mom and Dad then hang around until the larvae hatch, which can take up to four months.

Dwellers (paracoprid) neither roll nor burrow in the steaming pile of poo; they simply live in it. The female lays her eggs right there, and the entire development from egg to adult takes place inside the poo. Dwellers are significantly smaller than tunnellers and rollers.

In each group, we see a level of parental care that is a bit unusual for the insect world. I bring this up to soften your view of dung beetles. What they are doing is gross, but at least they are good parents.

Now, you might be asking yourself why all of this is important. *Is this just one more weird/creepy/gross thing Brenna is forcing me to learn?* Yes! I mean, no! No! How could you think that of me? This really is important. First, by burying animal poo, the beetles loosen the soil and improve nutrient recycling for plants to grow. The burying, mixing, and aeration of the soil also helps reduce methane release, so this action plays an important role in reducing greenhouse gas emissions from cattle farming—those gassy cows I told you about earlier in Chapter Three. Believe it or not, dung beetles are also aces at seed dispersal. Anything that the cows expel, they end up planting.

Dung beetles also help control fly populations. I haven't gotten too much into the flies, but the beetles and flies are competing for poo resources. Consider this: The average domestic cow produces as many as a dozen piles of poo a day. Each pile can produce up to 3,000 flies in a two-week period. In parts of Texas, dung beetles bury about 80% of cow poo. If they didn't, the flies would run rampant. Also, poo would cover and kill all the plants. Cows would have nothing to eat, and they'd all be driven to madness by the flies. Believe it or not, dung beetles save the US cattle industry an estimated $380 million every year by burying cow poo. It's no joke.

Just ask our friends Down Under. Australians learned this lesson the hard way when the Outback was nearly buried in cattle dung. Two hundred years ago, settlers introduced horses, sheep, and cattle to Australia—all grazing animals that were new to the native dung beetles. Australian dung beetles were raised on kangaroo poo and the like, and they refused to clean up after the exotic newcomers. Poo was beginning to pile up, and the fly infestations were untenable. Something had to be done. In the mid 1960s, the Australian government commissioned the Australian Dung Beetle Project. Some 23 species of dung beetle were imported from South America and Europe to take on the livestock poo. It was a weird thing to do, but the gambit proved successful. The imported beetles improved the quality and fertility of the livestock and knocked the fly population back by 90%.

Dung beetles play one other vital role. They are our canary in the coal mine. Scientists would phrase this differently. They would say that dung beetles are bioindicators. Through the actions of our poop-eating friends, we can clearly see the impacts of climate disturbances (extreme droughts and associated fires) and of human activities on biodiversity and ecosystem functioning (such as seed dispersal and nutrient cycling).

I'm going to make one last argument in support of dung beetles. To do their tireless work, at least one species of nocturnal African dung beetle, *Scarabaeus satyrus*, relies on celestial navigation to find its way.

With so many dung beetles vying for the same pile of poo, a beetle needs to make a quick getaway once he's got his ball ready. It's not easy to roll a poo ball in a straight line, especially if you're doing it backward with your hind legs. So, the first thing this little dude does is climb atop his poo sphere to orient himself.

The beetles use the entire Milky Way rather than individual stars to navigate straight lines from the dung pile to their homes. They discern varying gradients of brightness in the night sky, fixing on points of light to get those balls of dung to the right location. Researchers made this discovery after placing tiny hats on the dung beetles, effectively blocking their view of the heavens. Other dung beetles use celestial orientation as well. Diurnal dung beetles use the sun's position and scattered sunlight patterns to find their way.

Alas. Yet *one more* creature in this world that has better navigational abilities than me. I know my loved ones would say that bar is set pretty low . . . still. Can you imagine anything more amazing than this? These are beetles that eat poo, after all.

By now your brain is probably cramping. I've told you an awful lot in this chapter. You might need a rest. Unfortunately, there is no time for resting. In as much as I have already told you about how important insects are to our world, there's still a ton of important things I haven't told you yet. Press on, my friend, press on!

5.

AND EVERYTHING ELSE

UNCLE MILTON. THAT'S where the ant farm came from. (Not MY Uncle Milton; I don't have an Uncle Milton.) That's the name of the company: Uncle Milton. And he—or it, I guess I should say—sells the Genuine Brand Ant Farm. For a mere $31.80, you too could have a genuine Uncle Milton ant farm. It has, and I'm quoting, "the largest viewing area of any ant farm habitat!" Which is a fine feature, seeing how small ants actually are. It is "break resistant and escape proof!" Oh, I sure do hope so. It has a "tip-proof stand" and "clean tunneling sand." Better than those other models with tippy stands and dirty sand, for sure.

Uncle Milton's ant farm was first introduced to the world back in 1956. I think it's marketed to five-year-olds, but I don't care. This is important research. I need to observe ants directly.

Once you acquire your farm, you have to send off for the ants (harvester ants, genus *Pogonomyrmex*). There are a lot of rules about the sending of ants: They WON'T send them if the weather is too warm. Or too wet. Or too cold. Or the ants aren't ready yet. Or they think you are someone who will accidentally kill them. I made that last one up, but I think they should factor that in to their complex decision-making on who to send ants to and when.

Anyhow, the literature warns that you might have to wait between four and six weeks to receive your ants. It honestly takes about 7 minutes and 30 seconds to set up the farm. So it's good they warn you so you can busy yourself with other things while you wait for them to arrive. Myself, I thought I'd have plenty of time to prepare, which is why I bought the caterpillars at the same time. But that's another story.

Imagine my surprise a few nights later when I came home and found a pile of mail on the doorstep—among the bills and junk was a plain manila envelope. Not a PADDED envelope, mind you. No protective bubble wrapping anywhere to be seen. Just a flat, unassuming envelope with an Uncle Milton return address on it. It took a few minutes for my brain to process the name. Uncle Milton . . . Uncle . . . wait! That means ANTS! ANTS RIGHT NOW, not a month from now. Surely not? I opened the envelope carefully and two clouded-up, plastic tubes rolled out. They had so much funk in them, I had to hold them up to the light to confirm the moving shapes of live ants. Yikes! So. Not. Ready.

I called to the boys, suddenly grateful that Chuck was away and I wouldn't have to explain to him about my little insect project. The boys appeared in the kitchen to find me brandishing the nasty tubes.

"The ants have arrived!" I announced with flare. *Blink, blink.* No reaction. "I . . . uh . . . might need some help . . . you know, getting them . . . installed?" Liam nodded and sighed heavily.

Devin shook his head. "Why do you do this, Mom?"

"Do what?"

"You don't need ONE MORE THING to take care of!"

"This is research," I said meekly. "It's for Science."

Liam examined the scant paperwork from the envelope. "It says here we're supposed to put them in the fridge for 15 minutes. To calm them down before we install them."

And . . . 15 minutes later, I was standing at the stove. We turned on every light in the kitchen to help with visibility. The opening to the farm is a narrow slot on the top. It just so happened that the tube the ants come in is ever-so-slightly larger than the farm opening . . . Not very good design work there. Uncle Milton might have given that a bit more thought.

"CAUTION: Never handle ants directly." That's what the paperwork said. "Ants are easily excited and may bite or sting to defend themselves." Which explained what happened to Devin after I opened the first tube.

"After removing the tube of ants from the refrigerator, gently tap the tube so that they are at the bottom of the tube. Carefully open the tube and insert the end into the habitat as shown and tap/ shake the ants from the tube into your ant farm habitat."

Check! Tube tapped. Tube opened. Tube flipped down to connect with farm opening.

"If the ants seem sluggish at first, this is normal. They are tired and thirsty after their long journey through the mail."

Sluggish? No, even after 15 minutes cooling their jets in the fridge, we didn't see any signs of sluggishness. I'm no expert, but what we seemed to have witnessed was some sort of ant rage. It was definitely not a joyful *Are we there yet?* travel glee that we witnessed. It was more of an *I had to get up before dawn; I was put on a bad flight full of turbulence; they lost my luggage; and they gave me stale pretzels* kind of thing. The second I popped the top off, the ants surged forward. Devin instinctively put his hand near the opening to prevent any from falling to their death, only to be rewarded with a sharp bite.

He howled, which unnerved me to the point of shaking the tube with much more force than was necessary, sending even more ants spilling out of the tube. Most missed the entrance to the ant farm entirely and began spilling over the sides of the farm.

"Help!" I squealed. "Liam, help me!"

Liam stood behind me with his arms crossed. "And what, *exactly*, would you have me do?" he asked. "*WILL* them into the farm?"

"I don't know! I don't know," I squeaked. "But they're getting out!"

It took some quick thinking to invent ant-wrangling techniques on the fly. There were napkins and kitchen utensils involved. We did pretty well, I'd say. Only one loss of life. The others were corralled into their new home.

You're supposed to poke the sand with a little green stick provided by Uncle Milton to start some tunnels for them. Then, to be good hosts, you should offer them a little snack of sliced grapes. We did that. Grapes are naturally sweet, though, and all that sugar did nothing to calm the frantic buggers down. In between mouthfuls of grape, they seemed to be working through their plans to escape. Tunneling and climbing began immediately.

Each day thereafter, I observed what was happening in my ant farm. After several weeks, I realized the ants were slowing down. Some died. They seemed to move without purpose, directionless. What was going on here? I did some research and began to understand what was amiss. These ants had no queen. Like bees, ants are normally led by a queen. With the absence of such, the ants sort of devolve into lethargic chaos. I felt disappointed and a little angry about the whole thing. Why had Uncle Milton failed me? He hadn't, really. Most people who get ant farms tire of them fairly quickly, I'm guessing. This experience was never meant to last. It was supposed to entertain me.

INSECT INSPIRATION

Truth be told, we have relied on insects for a long time to entertain, enlighten, and inspire us, have we not? Look at our music, dance, literature, and art, and you'll see countless connections to insects.

We celebrate insects through song. Go way back. "El Grillo," a song about a singing cricket, was one of the first musical scores to be printed on Johannes Gutenberg's movable-type press in the 15th century. This hit number was written by Josquin des Prez, an Italian contemporary of Leonardo da Vinci, and includes the catchy phrase: "The cricket is a good singer/Who can hold a long note." Good stuff, Josquin. Good stuff. Not exactly dance music, but very popular in the day.

Early composers didn't need catchy lyrics, though, to capture the magic of insects. Consider Russian composer Nikolay Rimsky-Korsakov's masterpiece, "The Flight of the Bumblebee." Written as part of an opera in 1899, it depicts the moment when the tsar's son is transformed into a bee by a magic swan so he can fly off to meet his father, who thinks he's dead. You know, as is often the case with those Russian tsar types. The details aren't important. What is important is that in three and a half minutes, Rimsky-Korsakov captures perfectly the dizzying, frenetic movements of a bee darting hither and yon. If you don't know this piece of music, I will wait right here while you google it and listen.

Actually, that might take you awhile. This composition has fascinated people for so long, it has been arranged for almost every conceivable instrument. The original was written for a full orchestra, but if you pick an instrument, you'll find an arrangement for it. Cello? Yup. Clarinet? You betcha. Flute? Of course, yes. French horn. Check. Guitar. Uh-huh. There's even a thrash metal version of it from the 1990s by someone called the Great Kat. Go figure.

Less showy, but certainly no less complicated was Polish composer Frédéric Chopin's "The Bees" (Étude Op. 25, No. 2 in F Minor). And while you're in listening mode, don't forget "Song of the Flea" by Russian composer Modest Mussorgsky and the rather creepy-sounding piece "From the Diary of a Fly" by Hungarian composer Béla Viktor János Bartók. I'm sure there are many other examples. If you look hard enough, you'll find musical insect references in the strangest of places.

When Cab Calloway recorded "Call of the Jitterbug" in 1934 and popularized a new dance to go with it, the jitterbug probably had little to do with insects—but no matter. The jitterbug spread across the globe during World War II. By the late 1950s, it was arguably one of the most popular fast dances performed.

In the summer of 1957, Americans were in love with a singer named Jimmie Rodgers and his new hit, "Honeycomb." It isn't about beehives. It seems to be about a woman that Rodgers is hoping to marry, but . . . it's pretty catchy. That summer, it reached number one on both the Billboard Top 100 and the R&B Best Sellers charts.[1]

---

[1] Back in the day, these charts were how musicians measured their popularity and success.

I can't say if it was Rodgers's success that inspired other artists or if it was the insects themselves, but a lot of popular musicians have songs that pay homage to or at least make passing mention of insects. Who could say no to Godfather of Soul James Brown's "I Got Ants in My Pants (and I Want to Dance)"?[2] What about Van Morrison's "Tupelo Honey"? The Rolling Stones's "Shattered" references maggots. That's always good.

If I were to lecture you on cultural entomology, I would describe how insects have been used in our art for millennia. I would tell you about the religious and symbolic significance of insects historically reflected in our creative endeavors—as focal points in the delicate silkscreens of early Chinese and Japanese artists, as ever-present in the floral still lifes created by Jan van Huysum and other 18th-century Dutch realists, as eternal in the modern graphic art of Charley Harper, M. C. Escher, and Salvador Dalí.

CREATIVE GENIUSES?

That said, the man I really want to tell you about is Dr. Steven R. Kutcher. Dr. Kutcher is an American artist and entomologist. He doesn't just paint insects. He paints *with* insects. They are his living brushes.

He didn't start out doing such things. Dr. Kutcher is first and foremost an entomologist. He has a bachelor's degree in entomology from the University of California, Davis, and a master's degree in biology from the California State University, Long Beach. His formal studies focused on insect behavior, and this knowledge led him down an unusual career path.

---

[2] "I got ants in my pants / And I need to dance"—now there's some fine lyrics that really capture the nature of the ant.

Dr. Kutcher became an insect wrangler for Hollywood. Using his scholarly understanding of arthropod behaviors, Dr. Kutcher manipulated instinctive responses—such as sensitivity to light, air pressure, or gravity—to train insects and other arthropods to perform scripted tricks on cue.

Tricks? you say. *What sort of tricks?* Tricks like having a live wasp land harmlessly into the mouth of actor Roddy McDowall.[3] Or coaxing a spider to crawl across a room and into a slipper in *Arachnophobia*. Or having a cockroach crawl out of a shoe, walk up a bag of junk food and onto a surfing magazine, and then stop on a picture of a surfboard for *Race the Sun*. Or having a painted spider drop down on Toby Maguire's hand in *Spider-Man*.

This was back *before* the days of CGI.[4] Nowadays, if you want a cockroach to tap-dance across the table, you just call up an animator to render you one. Back in the day, if you wanted that effect, you had to find someone like Dr. Kutcher to train the cockroach to do that exact behavior. Dr. Kutcher's reign as wrangler started in the movie *Exorcist II: The Heretic*, where he used 3,000 African locusts. He also trained a fly to walk through ink and leave footprints behind for a television commercial for Steven Spielberg's *Amazing Stories*. At first, he didn't have the slightest idea how to do that, but he figured it out. And those projects launched a career—he has worked on more than 100 feature films with a bug in the storyline, including *Jurassic Park*. He worked in television as well—*CSI: NY*, *MacGyver*, and *The X Files*—not to mention

---

[3] McDowall was a British-born American actor who starred in *My Friend Flicka* and *Lassie Come Home* when he was a child. As an adult, he was best known for portraying Cornelius and Caesar in the original *Planet of the Apes* films. He was also in *The Poseidon Adventure* and *A Bug's Life*. Google him!

[4] Computer-generated images—a digital successor to stop-motion techniques that uses 3D models

commercials and, of all things, music videos. Yes, Dr. Kutcher wrangled a wasp for Michael Jackson, butterflies for Paula Abdul, and moths for Christina Aguilera. They call him the Bug Man of Hollywood.

I don't much care for that title because it sounds kind of flashy. Dr. Kutcher is anything but flashy. He is soft-spoken and gentle. He looks like a grandpa. *Why can't I have a grandpa like this?* you think to yourself. He's the sort of guy who you just want to hang out with because he knows stuff and because his house is full of aquariums with wondrous creatures living inside—tarantulas, beetles, cockroaches . . .

The early experience with the fly stayed with him over the years. He had always been fascinated by how insects move. But to see the remains of their movement captured on paper was really something else. It appealed to his inner artist as well as his entomological soul. So he decided one day to try it again, for the purpose of creating a piece of art.

To watch Dr. Kutcher[5] co-create with an insect artist is to be in the presence of something magical. Before he begins, he prepares his canvas by wetting the paper. Dr. Kutcher has in his mind the sort of art he wants to create. He will choose the colors, the texture, and the composition of the piece. With this vision, he chooses his co-artist carefully. Working with each insect is like using a different kind of brush because each insect has a unique way of moving. Dr. Kutcher knows, for example, that cockroaches drag their tummies a bit when they walk. If he applies paint to their feet *and* tums, it will create a certain effect. He knows that

---

[5] Go here to see his work: bugartbysteven.com. Some of his work is also for sale.

with some types of moths, the tips of their wings will brush across the paper, so he paints their feet and wing tips.

As he makes his selection, Dr. Kutcher gently lifts his co-artist and paints its feet with a water-based, nontoxic paint. Then he sets the artist down on the paper, and the work begins. A painting might take more than an hour. It could take days. It might require the work of several artists. Dr. Kutcher may turn the paper as the artist works, creating a paint trail of spirals and circles. Dr. Kutcher works in complete communion with the insect. The result is an astonishing piece of original art. It is, in many ways, a small masterpiece—a joining of species to create something pleasing to the eye and a wonder for the soul.

What happens next might surprise you. Or maybe not. Dr.

Kutcher gently gathers his co-artist up and places it in the sink. He turns on a small trickle of water and allows the insect to walk around the basement, cleaning off its feet. Dr. Kutcher uses a spray bottle to remove any residual paint, then places the artist in a shallow bowl that has been lined with a wet paper towel. If the insect has even a small trace of paint still on its feet or body, Dr. Kutcher will rinse his friend again. Once sure that the artist is paint-free, Dr. Kutcher returns it to its habitat with an offering of food—a thanks for all its hard work. It's difficult to fully describe the care that goes into this final ritual. You really are left in awe of it.

SUPERHEROES AND VILLAINS

What other ways do insects inspire us? Entire tomes could be written about the use of insects in our storytelling traditions. If I asked you to a name an important literary work that centered around insects, you might be quick to cite Franz Kafka's *The Metamorphosis*. Gregor Samsa wakes up one morning to find himself transformed into a huge, icky bug.

The Metamorphosis is certainly an important work. It's been unnerving high school kids in English classes for generations. But if I'm honest with you—and you know that I do try to be— when I think of the intersection of insects and literary genius, I'm more apt to think of Eric Carle's *The Very Hungry Caterpillar*. What a magnificent book! That poor little fella . . . getting a tummy ache after eating so much. It's probably that last piece of salami that puts him over the edge. It all works out in the end, doesn't it?

Have you gotten to know Redd from Kelly Murphy's wonderful *The Boll Weevil Ball* or Jerry Bee from *Bee-Wigged* by the great

Cece Bell? Such books! Roald Dahl's *James and the Giant Peach* and George Selden's *The Cricket in Times Square* both have standout insect characters.

While we're on the subject of literary greatness, let us not forget the scores of insect-inspired heroes and villains that Marvel Comics, DC Comics, and others have dreamed up over the years. I'm sure you can rattle off quite a few.

No, no. Not Spider-Man. Don't go there. Spiders . . . aren't insects . . . You keep forgetting. So while Spidey is one of the greatest comic book superheroes EVER, we can't put him on our insect list. That also goes for Scorpion, Black Widow, and The Tick. Good characters but not insects.

Ant-Man, though. He's totally fair game, right? And where there's Ant-Man, there's also The Wasp and their nemesis Yellow-jacket.[6] Insect-inspired characters tend to be more villainous than heroic, sad to say. They get their powers from being mutants or from having a super suit or from being bitten by some altered insect. Many of their superpowers are derived from real insect powers like flight, strength, speed, stinging abilities, wall climbing, or enhanced sensory perception (hearing or sight).

Mantis, who you may have most recently seen in the *Guardians of the Galaxy* films, is a great example. She has enhanced strength and superior vision, but her antennae also help her perceive things that others can't and allow her to be an empath.

Most superheroes and super villains have complicated back-stories. And some characters are so loved and have been around so long that they get reimagined and reinvented. Blue Beetle is one with incredible staying power. His identity has been used by

---

[6] Although, she does repent at some point and turn into a Good Guy.

three different heroes since 1939. His first incarnation came in the form of archaeologist Dan Garrett, who unearthed a scarab made by an alien species. The scarab had been used to imprison an evil mummified pharaoh, but when Garrett utters the mystical words "Kaji Dha!"[7] with gusto, he is imbued with incredible powers, including super strength and super vision, flight, and the ability to generate energy blasts. Pretty keen, huh? Unfortunately, Garrett meets an untimely death while investigating an evil robot army, and so it is left to his protégé Ted Kord to step into the role. Only Kord doesn't know the magic words, so he can never quite catch the scarab's power. He has to improvise, relying on his own manly strength, pure heart, and derring-do to defeat evil and live up to the Blue Beetle's name. Until he, too, meets with an untimely death. Fans were pretty upset about that, so the mantle had to be passed to another. Jaime Reyes, a Texas-based teenager, figures out that the scarab can transform into an exoskeleton suit that's full of alien weapons, and the story line continued from there.

A number of insect-inspired superheroes lack superpowers, but they have neat-o costumes or cool inventions that they use to fight crime. Take the Green Hornet. He's just a regular bloke. A newspaper publisher by day but masked crime-fighter by night. He even wears a suit. Not a superhero suit but a *suit* suit. He prowls the streets clad in a long green overcoat, gloves, green fedora hat, and green mask. The only thing he's got going for him besides a sense of righteousness is a gun that fires knockout gas and another one called the Hornet's Sting that sends out electric shocks. Kinda

_____

[7] Sorry. I'm at a total loss for what this might translate to, but it works for him.

handy, I should think. Green Hornet's been around since 1936; he's pretty old-school.

Here's one you've probably never heard off: the Red Bee. Richard Raleigh is an assistant district attorney who grows tired of seeing criminals get off on technicalities. He decides to do something about it. Unfortunately, Raleigh's only true skill outside of lawyering is beekeeping, so he has to find a way to use that to his advantage. Decked out in stylish red-and-yellow-striped tights, our man Raleigh takes on Bad Guys with a squadron of *trained bees*. His one secret weapon is a bee named Michael who hangs out in a secret compartment in Raleigh's belt buckle. When things get dicey, Raleigh springs open his buckle to release Michael, who, unlike normal bees, can sting more than once without dying.[8] As this was the 1940s, Michael would annoy and bedevil Bad Guys— who were most often Nazis—by repeatedly stinging them until they relented.

Not all insect-inspired characters were heroes, though. Firefly (Garfield Lynns) started off as a criminal who played with lighting and visual effects to successfully pull off robberies. When that got old, writers eventually turned him into a pyromaniac who sets things alight with wild abandon. No real insect powers, of course, but he did wear a snazzy, fireproof battle suit and usually arrived at crime scenes packing a flamethrower.

Killer Moth was a bit more buggy. He was a small-time criminal named Drury Walker, who adopted the persona of a millionaire philanthropist so he could sidle up to and befriend Bruce

---

[8] Obviously a great deal of poetic license was taken here. Male bees don't have stingers. I also can't imagine Michael would care for being stuffed away in a belt buckle for prolonged periods of time, even if it did mean he'd have his chance later to torment Nazis.

Wayne, aka Batman. Under his Killer Moth persona, he then went mano a mano with Batman using advanced weaponry like an infrared Moth-Signal. He also tooled around in his Mothmobile. No superpowers, but he had great bug tech, including a cocoon gun that fired sticky threads. In one plotline, he metamorphized into a giant part-moth creature with wings, an exoskeleton, and acid spit. That was something. There are many other great examples of insect heroes and villains, of course, but you get the general idea. Insects were either reviled or venerated in comic books for their unique capabilities.

TRENDSETTERS

When we aren't turning to insects for literary inspiration, we're looking to them to maintain our sense of fashion and style through the silk trade, dyes, and beauty products.

For thousands of years, we've relied on the giant silk moth (*Bombyx mori*) to weave a cocoon for itself as it transitions from pupa to adult. And for thousands of years, we've confiscated these cocoons and repurposed their silken threads. Silk production, or sericulture, is a complicated process. It starts with a female caterpillar laying her eggs—as many as 500 at a time. These eggs hatch into larvae, called silkworms. The silkworms are only fed mulberry leaves to encourage their growth.

After about six weeks of near-constant feeding, they've grown to be about 7.6 centimeters (3 inches) long and are ready to start spinning their cocoons. Silkworm silk is a continuous thread secreted by two salivary glands in the head of each worm. The worm repeatedly rotates its body in a figure-eight movement to spew the silk around itself, probably in the neighborhood of about

300,000 times—although I'm not sure who did the counting on this.

Only a single strand of silk is produced, measuring about 100 meters (328 feet) long. It is held together by a type of natural gum called sericin.[9] It takes about a week for the larva to spin and fully encase itself. Then we come along to harvest it. We do so by boiling it in hot water, thereby killing the pupa[10] and dissolving the gum that holds the cocoon together. Now the single strands can be unspun and combined with others to form a silken thread. One thread is made up of about 48 individual silk strands. The threads are later woven into a fabric.

We have harvested silk from this one species for so long, it is thought to be entirely extinct in the wild.[11] China is the world's single biggest producer and chief supplier of silk to the world, and India's a close second. More than 100,000 metric tons of silk are produced annually across some 60 countries. Think of how many silkworms that is.

OTHER PRODUCTS

We harvest other things too. You probably don't think too often about shellac. I know I don't. Heck, you might not even know what it is. Shellac is a resin secreted by a tiny red insect called the lac bug (*Kerria lacca*). Swarms of these insects feed on certain trees

---

[9] It takes around 2,500 silkworms to produce 1 pound of raw silk.

[10] Which people later eat

[11] These are not the only caterpillars that produce silk, of course. Nor are they the only animal to do so. Spider silk is extremely strong, but people haven't figured out how to harvest silk from captive spiders in an economical way. Oddly enough, during the Second World War, spiders were employed by the US government to spin silk to be used to make cross hairs in the gunsights of US Army weapons.

in the forests of India and Thailand. Lac bugs only live for about six months, so they make the most of their time here on Earth. They settle themselves comfortably on tree branches to gluttonously suck on tree sap.[12] That's priority number one. Priority number two is to produce as many offspring as possible. Each female lac bug might lay 1,000 eggs in her lifetime. All this eating and mating can take a toll on a lac bug's system, hence the secretion of the resin onto the tree branches. The tree sap, of course, is chemically altered, having run its course through a lac bug's digestive tract. Upon contact with the air, the excretion forms a hard, shell-like covering over the tree branch.

Later, people come along and cut these branches for harvesting and processing. The shellac is sold as dried flakes. The flakes are later dissolved in ethyl alcohol to make a wood finish. When applied, liquid shellac dries to a high-gloss sheen. Shellac is used in a lot of other ways too. Combined with wax, it can be a preservative and shine-inducer for citrus fruits. It can be used in certain dental technologies. It can be used in bike handlebars. It can be used in fireworks, in watchmaking, and for archery arrows. And probably strangest of all—it's used in Jelly Belly jelly beans, in combination with beeswax, to give them their final buff and polish. Shellac: Something you never think about but something that is in all kinds of things, and which we rely on insects for. To make a small quantity of shellac—say 1 kilogram (2.2 pounds)—you might need the services of as many as 300,000 lac bugs. Now that I've told you this, you're likely to never forget it because it is such a random thing to know.

Here's something a little more obvious in your daily life:

---

[12] Indigenous people call this the feast of death.

CI 75470. Ring any bells? No? How about cochineal extract or carmine? Still nothing? Check your care and content labels on products you use for "natural red 4." This dye was traditionally used to color fabric,[13] but is now more commonly used to dye food and makeup products red or pink. The dye comes from a chubby little scale insect that feeds on the prickly pear cactus. Female cochineal insects (*Dactylopius coccus*), native to Mexico, have no legs or wings and are little more than a bag of guts and eggs covered in a white wax that constantly sucks up plant sap. The red carminic acid, which the insect produces from eating the cactus pads, evolved to protect it from predator ants. When the bugs are crushed, their vibrant red dye is released. It takes the magnitude of some 70,000 cochineal insects to produce less than 0.5 kilograms (1 pound) of dye.

Starbucks made the news a number of years ago when it announced that it would no longer use natural red 4 in its Strawberry Crème Frappuccinos because, frankly, it was grossing everyone out. Most of those tender-hearted Frappuccino drinkers probably had no idea how ubiquitous that dye really is. It's in your raspberry yogurt, soft drinks, fruit drinks, energy drinks, candy, Popsicles, chewing gum, canned fruit, canned soups, applesauce, maraschino cherries, artificial crab meat, ketchup . . . Cochineal extract is such a handy dye, it's also used in many, many hair- and skin-care products, lipsticks, face powders, rouges, and blushes.

Jazzing up our beauty products with crushed-up scale bugs is nothing, though. We also rely heavily on honey bee products to maintain our beauty. One of the kookier products I've run across

---

[13] It was all the rage among the Aztec and Maya peoples of North and Central America as early as the second century BC and brought to the rest of the world by the Spanish. Before long, you couldn't be a proper Catholic cardinal or English redcoat without this dye.

uses bee venom to give you pouty lips. I have never felt the need to *have* pouty lips, but for many others, pouty lips appear to be highly sought after.

Lip plumpers work on the principle that applying an irritant to the lips will puff them up. A number of ingredients can be used to cause this effect: cinnamon, wintergreen, cayenne pepper, caffeine, ginger, or (you guessed it) bee venom. These awful things loaded into your lip balm or lip gloss boost blood flow to the lips, leading to mild swelling and redness. Which doesn't seem like a desirable state to me, but what do I know about it?

One product goes so far as to promise a new technology that allows collecting bee venom without harming the bees. I seriously doubt that. What do they do? *Milk* the bees for their venom?

INSECTS IN MEDICINE

We *do* know that bee venom has been linked to other medical benefits. Some research suggests that bee venom may help ease joint pain, swelling, and stiffness in people suffering from arthritis. It's believed that bee venom slows the production of interleukin-1, a compound that helps fuel arthritic pain and inflammation. People who rely on bee venom as a therapy must willfully allow themselves to be stung—in some extreme cases, 80 times or more a day. I tend to hit the Aleve bottle pretty hard to deal with my arthritic pain because I'm really not a fan of getting stung.

To this point, though, there's an entire branch of alternative medicine called apitherapy that uses honey bee products, including bee venom, pollen, propolis, royal jelly, and honey. Of course, honey has long been thought to have curative powers. What do we drink when our throats are scratchy or when we feel poorly?

Hot tea with honey. The honey soothes our throats and makes us feel better. Everybody knows that. But I didn't always know the value of honey's antimicrobial properties. Nor did I ever know that honey could be used in wound treatment. I found that out one day a long time ago while volunteering at a place called City Wildlife. It's DC's rehabilitation center to help injured and orphaned wild animals. We once treated a wounded opossum by putting honey on its injuries.

SUTURE ANTS

In some cases, we resort not to the by-product of an insect but to the insects themselves to help us medically. Not that I would ever recommend this treatment if it could be avoided—but . . . say you're an explorer dithering about Central or South America, and you manage to accidentally impale yourself on a pointy branch. *Oops!* you think. *My bad.* Rather than bleed out in the rain forest, you might want to suture up that nasty cut. But how? You seem to be a bit short on medical supplies. *Eciton burchellii* to the rescue! Equipped with sharp, sickle-shaped mandibles, the suturing army ant might be your best hope for closing that wound. This definitely won't be a comfortable exchange between you and the ant. When grasped from behind the head, the ant will open its jaws wide. By placing one mandible on each side of the cut, the squeezed ant will instinctively perform its normal defensive behavior—biting into your skin. To keep its head locked in place around your wound, you have to sever the head from the body. The ant's locked jaws create a suture that can hold for days.

That's a rough bit of emergency medicine right there, but I think I'd prefer it over other forms of insect medicine. I'm thinking

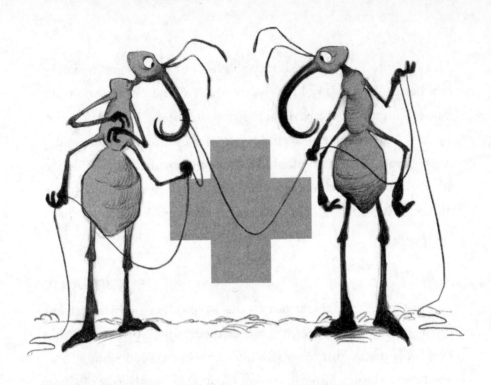

specifically of having a doctor pour live maggots into an open wound.

MAGGOT THERAPY (YES, REALLY)

Yes, this is an actual therapy. It is a remarkably effective therapy but not a new one. For centuries, the beneficial effects of maggots in wounds were noticed by military surgeons. It feels counter-intuitive, but injured soldiers abandoned on the battlefield seemed to fare better and their wounds healed faster when those wounds attracted and were infested by maggots. Incredibly, the maggots would eat the dead tissue and clean the wound. Soldiers showed lower rates of infection and faster healing times.

An orthopedic surgeon named Dr. William Baer observed this during his service in World War I. He wanted to pursue the idea by intentionally applying live, medical-grade fly larvae to wounds. After the war, he was the first to systematically apply maggots to nonhealing wounds while at Johns Hopkins and Children's Hospital in Baltimore, Maryland. The treatment was successful in more than 100 patients. Dr. Baer published his studies in 1929, and within five years, his groundbreaking work had inspired more than a thousand American, Canadian, and European surgeons to begin using maggots in their practices.

While promising, this new therapy did have some drawbacks. It was expensive. It was hard to acquire medicinal maggots. And it was messy. The bandages made to secure the maggots to a wound often malfunctioned, which resulted in maggot escapees. By the 1940s, maggot use was largely abandoned in favor of widespread antibiotic use.

Nonhealing wounds continued to be a problem, though, especially in diabetic patients. By the end of the 1980s, doctors saw a large increase in pressure ulcers and diabetic foot ulcers. If these sound dire, it's because they are. We used to call pressure ulcers bedsores. They are injuries to skin and underlying tissue caused by prolonged pressure on the skin. Skin that covers bony areas of the body (such as the heels, ankles, hips, and tailbone) are most susceptible. People at risk often have medical conditions that limit their ability to change positions or cause them to spend most of their time in a bed or chair.

Diabetic foot ulcers can form due to a combination of factors, such as lack of feeling in the foot, poor circulation, or irritation caused by friction or pressure. Patients who have diabetes for many years can develop neuropathy—a form of nerve damage. The nerve damage can happen without pain, so a person may not even be aware of the problem.

Today, the annual cost of management for these types of wounds exceeds $20 billion, not including the loss of two million workdays. Diabetic foot ulcers alone are so common that they account for more than 1.5 million cases and at least 70,000 amputations every year.

Why the steep increase in cases? People are living longer. Conditions that would have been fatal 40 years ago are now medically managed into chronic conditions. They aren't going away, but they aren't killing us like they did either. Plus our arsenal of antibiotics is shrinking because of the growth in antibiotic-resistant microbes.

Desperate to find relief for their patients, a number of doctors turned to older therapies, which has led to our reacquaintance with the maggot. In 2004, the FDA cleared the maggots from the common green bottle fly (*Lucilia sericata*) as a medical device in the United States for the treatment of these types of wounds and several others. The maggots' job is to debride a wound (which means to remove damaged tissue), disinfect it, and promote healing.

Here's how it works: The maggots are chemically disinfected to make them germ-free. They are then applied to an open wound at a dose of 5 to 10 larvae per square centimeter of wound surface area. They are covered by a special bandage, which allows air to pass through the bandage but doesn't allow the maggots to wander beyond the wound unescorted.[14]

What's going on underneath that bandage? The maggots eat away all the dead tissue in the wound, and they do it with a greater level of precision than even the best surgeon could hope for. What's more, as they're eating away all the yuck in the wound, they're

---

[14] Apparently, these bandages are also designed to cut down on the unnerving tickling sensation that a patient can feel while being feasted upon by live maggots.

excreting enzymes believed to have broad-spectrum antimicrobial activity, which will promote healing.

The bandage is kept in place for 48–72 hours. At that point, the little buggers have had their fill of eating the rotten tissue. While hiding out in the wound, the maggots will have molted *twice*, increasing in length from 1 millimeter (0.03 inches) to 10 millimeters (0.39 inches).

Ugh. So I think we can agree: SUPER GROSS. However, it's a very effective therapy. Medicinal maggots are widely acknowledged to be safe, effective, and now relatively inexpensive for this kind of wound treatment. In published studies, maggot therapy is associated with limb salvage in up to 60% of patients scheduled for amputation. If you are one of those patients, this is nothing short of miraculous.

The truth is, we've relied on insects to help us medically for a really long time. Think back to all the genetics studies that were made possible by *Drosophila melanogaster*, the fruit fly. If you want to do some interesting research, look into their last 100 years. They've played a starring role in biomedical research, revealing fundamental principles of genetics and development, illuminating human health and disease, and earning at least six scientists Nobel Prizes.

INSPIRATION FOR INNOVATIONS

Insects solve other problems for us too. They inspire new designs and innovative technologies. African termites led us to the way of a new office structure. It's the Eastgate Center in Harare, Zimbabwe. This large office building and shopping mall was built on the same design principles that termites use when constructing their mounds. Scientists observed that these mounds stay remarkably cool inside, even in blistering heat. The insects accomplish this by creating their own air-conditioning systems that circulate hot and cool air between the mound and the outside.

Mick Pearce, the building's architect, wanted to recreate the way of the termite on a larger scale. Eastgate uses a system of passive cooling—it stores up heat during the day and vents it at night. At the start of a day, the building is cool. As people use the building, heat is generated. The sun shines in. That heat is absorbed by the fabric of the building, which has a high heat capacity. By evening, the temperatures outside begin to drop. The warm internal air naturally rises because it is less dense. Aided by fans, it is vented through chimneys. This draws in denser, cooler air at the bottom of the building. As a result, Eastgate naturally warms and cools itself and uses 90% less energy than a same-size building that uses forced air heating and cooling. This efficiency is better for the environment and results in lower heating and air conditioning costs, which translate into lower rental prices for tenants. Everyone wins.

Just by paying attention to insects, we can learn valuable lessons. We owe a lot to Ki-Tae Pak, a student at Seoul National University of Technology, for his observations and later application of the unique behaviors of the head-stander beetle (*Onymacris unguicularis*). This beetle thrives in the hostile, arid Namib Desert of southern Africa. How? The head-stander is a fog-basker.

Where there's water, there's life, and the Namib Desert is surprisingly full of life thanks to an odd water source: fog. A cold ocean current—the Benguela Current—brings icy waters from the depths of the Atlantic Ocean to South Africa's and Namibia's western coasts. The Benguela interacts with warm moisture above the sea surface, which cools and forms dew and then fog. Southwestern winds then push the fog inland across the desert.

The head-stander shinnies up the nearest sand dune in the early morning and tips its head forward, pointing its bum skyward. Fog condenses on its backside and water droplets run down the ridges of its body toward its mouth for a drink. Pak's Dew Bank Bottle imitates the beetle's water-collection system.

Left out overnight, the stainless-steel Dew Bank Bottle takes advantage of the night's lower temperatures, which cool the metal. In the morning, when the surrounding air begins to warm, water droplets condense on the cool surface of the bottle. The water is channeled down ridges in the surface of the bank to an enclosed, circular holding chamber. The bank can gather enough moisture for a full glass of water. Such a thing will be of enormous value to people living in places where fresh drinking water is scarce, and the potential for this solution to be scaled-up for greater water collection can now be explored.

When Resilient Technologies and the University of Wisconsin–Madison's Polymer Engineering Center decided to reinvent the wheel, that wasn't a figure of speech. I've never given any serious thought to a military vehicle's tires before, but during combat, a vehicle's wheels can be its Achilles heel. You can armor a vehicle, but you can't really armor its tires. Tires are susceptible to bullets, debris, and even rocks that can cause punctures. A puncture can ground a mission to a halt, putting the vehicle and the people it's carrying at risk.

So these folks at Resilient and W–M created something new: It's an airless tire that's bulletproof, can withstand the force of explosions and gunfire, and can hold up on extreme terrain. Guess what it's modeled after? Go ahead, guess. You'll never get it, so let me just tell you: honeycomb. That's right! It's back to the bees for inspiration!

The technology is called non-pneumatic tire (NPT), and it is made of a rubber tread supported by a lattice of plastic hexagons with a central metal bolting hub, which is car talk for it looks like the cells of a beehive.

The walls of the honeycomb structure spread the load more evenly and permit better heat dissipation—also things I didn't know tires needed, but there you go. Aside from the obvious benefit of never going flat, these tires are also lighter than standard ones and there's no need to carry a spare or check the pressure with that little stick thing my dad was always showing me how to use. If it keeps our men and women in combat zones safe, I'm all for it.

NEW RECRUITS, NEW ROBOTS

The US military might soon be recruiting insects as well. DARPA—the Defense Advanced Research Projects Agency—has an entire program devoted to cyborg insect spies. DARPA is the wing of the Department of Defense that's responsible for developing emerging technologies for military use. Using implants, researchers have shown that it's possible to stimulate insect brains and control them in flight. Eventually such insects could be used in the field to gain access to areas not easily reachable by humans or robots. There's some other stuff I need to tell you about, but not *here*, not *now*. I'll have to debrief you *later*, after you've passed your security clearance.

We *can* talk about robots, though. Do you remember HEX-BUGs? These are little toy robots based on arthropods. The first one was called the beetle, but it was the tiny, vibrating nano that really took the world by storm. They are finger-thin with soft, rubberized feet. They come in many colors, and you can buy entire habitats for them to skitter around in. Before long, the company that made them, Innovation First, had created all sorts of clever variations, including mini bots that battle each other.

It probably won't surprise you to learn that a number of serious roboticists have created other robots based on insect design. Insects are engineering wonders—their size . . . the way they walk, climb, fly . . . their incredible strength . . . so many enviable attributes and abilities. There are a lot of robots in development right now. We'll look at four, which are based on the bee, the termite, the ant, and the cockroach.

When roboticists at Harvard's Wyss Institute for Biologically Inspired Engineering set out to develop a tiny, flying robot, their initial inspiration was a fly not a bee. They knew they wanted the robot to fly, but because it is absurdly small—about half the size of a paper clip and less than one-tenth of a gram in weight—it could not be built with rotary motors, gears, nuts, and bolts. Those things simply would not be viable at that scale. They were too complex and too heavy.

So they took their inspiration from another source: pop-up books. They developed a technique to cut the parts of the robot from flat sheets that had been layered. Then they folded the sheets like a piece of origami. Glue was used to hold everything in place.

They realized that at micro scale, any amount of wind could have a dramatic impact on the robot's flight. It had to have rapid reaction response capabilities. To that end, they built wings with artificial muscles imbedded with thin ceramic strips that contract when an electric current passes through it. Plastic hinges allow the wings to rotate independently from each other. To achieve flight, the wings must beat 120 times per second. Oh sure, it sounds *simple* now as I explain it to you, but just to get this thing airborne took more than a decade of work. Its 3-centimeter (1.2-inch) wingspan makes the RoboBee the smallest man-made device modeled on an insect to achieve flight. The RoboBee is remarkable, and the applications for its use are far reaching—crop pollination, search and rescue missions, and surveillance as well as high-resolution weather, climate, and environmental monitoring.

Now, it just so happens that Harvard is a friend of the insect. The same lab that brought us the RoboBee also created a robot based on the termite. Termites, as we know, are fantastic builders. They create some of the largest structures by any nonhuman animal on the planet. A termite mound is an intricate structure filled with tunnels, chambers, galleries, and staircases. To create such things, termites have to move a lot of raw materials. In the course of a year, 4.9 kilograms (11 pounds) of termites can move about 165 kilograms (364 pounds) of dirt and 1,497 kilograms (3,300 pounds) of water. By its little lonesome, a single termite has a limited range of things it can do: crawl, turn, dig, and stack a mud ball. However, working together, they can build skyscrapers.

If that wasn't impressive enough, termites seem able to complete their tasks without use of a foreman. This is something called stigmergy. Instead of directly communicating with each other—like, "Are you going to put that there?"—they instead read their environment and work from there.

We don't quite understand how they do this, but it might have something to do with spit. Most termites don't have eyes, so they aren't looking at what their fellow termites are creating so they can follow suit. No, instead they seem to communicate through pheromones. The trigger to act may be found in some sort of chemical in their saliva. If one termite picks up a mud ball in its mouth, the mud ball gets a bit spitty. Other termites, triggered by the smell of the saliva, start stacking mud balls too. This magnifies the original spit signal and before you know it, you have an entire mud ball wall built.

It's just this sort of autonomous building that roboticists are interested in. The folks at Harvard's Wyss Institute created tissue-box-size robots as part of their TERMES project that could navigate and move building blocks in their environment to build structures. Each robot was created to perform several simple behaviors: move forward, move backward, turn, move up or down a step, and move while carrying a brick. To sense their environment, the robots were outfitted with a number of sensors. Each robot is tricked out with whegs—a combination of wheels and legs that enable them to roll along but also manage stairs. And each robot has a spring-loaded gripper to lift, carry, and place blocks.

By following a sequence of programmed steps, each robot can build a structure. What's more, if you have a group of TERMES robots, they will collectively build the same structure, without being told what to do or without communicating directly with each other. If one robot gets in another's way, it just waits until it stops sensing the robot, then continues building.

Why would such a robot be of interest? The Harvard team foresaw several uses—everything from moving sandbags and building levees in flood zones to building a base for human habitation on Mars.

The Harvard folks aren't the only ones interested in group behavior for their insect-inspired robots. James McLurkin is a senior hardware engineer for Google, but back in the day, when he was a graduate student, McLurkin built robots based on ants.

Like the TERMES robots, McLurkin was looking to create groups of robots with populations in the 10 to 10,000 range that could work together to perform tasks. His tiny, square robots had small internal computers that allowed the robots to communicate, cooperate, and move relative to each other. Some behaviors are simple, like following the leader, clumping together, or dispersing. Some are more complex, like performing a dynamic task.

Robots are often created to complete jobs that meet one of the three Ds: dangerous, dirty, or dull. With the ability to travel 25 centimeters (10 inches) per second, these ant-inspired robots were created with the goal of exploration, search and rescue, recovery, mapping, and surveillance.

Meanwhile, biologists at the University of California, Berkeley, were looking to cockroaches for inspiration. The Dynamic Autonomous Sprawled Hexapod, or DASH, is a cheap, featherweight, six-legged wonder—10 centimeters (4 inches) long and 5 centimeters (2 inches) wide. It's made from cardboard laminated with flexible polymer using a 3D printer.

Like a cockroach, it can move steadily across multiple surfaces at high speed—15 body lengths per second. DASH can climb over obstacles taller than its own body height and inclines up to 17 degrees. DASH has been modified with interchangeable feet that include claws or magnets, allowing the robotic cockroach to climb vertically. Recently, the robot has been given wings and the ability to fly as fast as 1.3 meters (4 feet) per second.

It can also do something special that cockroaches can do. When DASH gets to the edge of a surface, such as a table, it can hook the edge with its rear feet and swing around 180 degrees to land on the underside of the table and keep going. When DASH is dropped from a height, it can recover from that too—on account of its flexible skeleton. Such a handy, resilient robot could prove to be a particularly valuable ally in the aftermath of earthquakes.

COMPLEX PROBLEM SOLVERS

Insects have inspired robots, but they've also inspired entire solution systems for complex problems. I'm talking about trash. I'm talking about trash in China and what the humble cockroach might be able to do about that. China generates at least 60 million tons of kitchen waste annually. Most of it gets processed through fermentation, which is expensive and somewhat inefficient. It also pollutes the environment. If only a better way existed. Enter Li Yanrong. He left his corporate job years ago to start a new business. If you visit him in Jinan in the Shandong province, he might let you tour his facility and meet the more than 300 tons of hungry American cockroaches (*Periplaneta americana*) he keeps there. Li's roach farm is capable of consuming 60 tons of food waste *every day*.

The process is absurdly simple. Food waste is collected from restaurants and delivered to the warehouse at night. The food is separated from any packaging like plastics, glass, or metal. Then it's blended into a mush, which is channeled through a pipeline to feed the cockroaches.

The idea came to him after watching educational videos about cockroaches and their eating habits with his daughter. He knew that most food waste was being buried in landfills and wondered how much of that food could be consumed by cockroaches instead. He experimented with a fish tank full of cockroaches in his living room and soon discovered that cockroaches are none too picky about their diets.

Armed with automated ventilation, feeding, cleaning systems, and good intentions, he started his new business in 2011. He chose the American cockroach because it is the largest, most common, and has the longest life span. They live on his farm for an average of 300 days.

Even in death, though, they continue to serve. After they die, they are collected and ground into a powder that is put into animal feed. Not surprising, cockroaches are rich in protein. According to statistics from the Jinan Food Quality Supervision and Testing Center under the Ministry of Agriculture and Rural Affairs, chickens fed with cockroach powder can survive without antibiotics, the fat content of their meat is lower than that of rabbit meat, and their selenium[15] content is 1.8 times higher than that of ordinary chickens.

But back to those billions of cockroaches. What if cockroaches escape? Fair question. That would scare most people, but not to worry, Li has thought of everything. The warehouse has been designed with multiple sealing measures like water curtains at every entrance. The fishponds that surround the warehouse are another line of defense. So far, no incidents have been reported.

I have never met Li Yanrong, but he is my personal hero. This

---

[15] Selenium is a trace element that is nutritionally essential for chickens. It plays a critical role in reproductive success, egg production, and hatchability.

man is single-handedly solving a formerly insurmountable problem, and he's using one of the most reviled insects to do it. You can tell that I have a great deal of respect for the insect world. That said, I would be remiss if I didn't address the ways in which insects suck and make life miserable for people too. In our next chapter, that's what we'll tackle.

6.

HOW INSECTS SUCK

"CICADAS. WHY?"

That's Chuck speaking. He wants to know why cicadas are on this Earth.

It's probably not a question you've ever pondered, but if you live in Washington, DC, it's something you ask yourself every 17 years. We've been poised for weeks now, expecting Brood X. That's not the letter X. That's X as in the Roman numeral for 10. Here, it denotes the largest of 15 groups of periodical cicadas that regularly appear throughout the eastern United States.

When they last appeared, I was pregnant with Devin, and we were still living on Capitol Hill. Few homes on the Hill have much of a yard, so while I remember there being a lot of media coverage about the cicadas coming, I don't remember actually seeing a single cicada. This time would be different! During the

interim, we moved to a home situated across from Rock Creek Park, and it has a yard. We're sitting at ground zero now.

You may not have ever experienced this East Coast phenomenon,[1] but you've probably seen a cicada in your lifetime. Annual cicadas, also known as dog-day cicadas, have dark green bodies and green veined wings. They show up, as their name suggests, every summer. No big deal. Periodical cicadas are something else. They have bulging, red eyes; black bodies; and reddish-orangish-veined wings and legs. They're sort of blocky with biggish heads and are about as long as your pinky.

These things live most of their lives buried alive, sucking on xylem from tree roots. At the 17-year mark,[2] when the soil temperature reaches about 17.7°C (64°F), they tunnel up to the surface. Once you spot finger-size holes in the ground near the base of trees, you know to start looking for them. They molt one last time (leaving behind a paper-thin, creepy-looking husk of an exoskeleton), then spend the rest of their six-week, adult lives eating and mating.

Why do they do this? I don't know. I'm sorry. It's all rather mysterious. It's been a well-documented occurrence since the 1700s. And when they do arrive, they do so in numbers—as many as 1.5 million individuals per acre. So we're talking trillions. With a *T*.

Every media outlet has been hyping the coming invasion. Naturally, with this many insects crawling out of the ground at the

---

[1] For the 2021 emergence, some 35 million Americans had the opportunity to interact with these critters as they covered a geographical range stretching from Tennessee to New York.

[2] Some species of cicadas resurface after 13 years.

same time, people are bound to freak out. I'm wondering if Earth will stop spinning on its axis.

When the first cicadas arrive, though, I'm fairly oblivious. It's the crack of dawn. Zorro, our pug, and I are out for our first stroll of the day. He is very focused, checking the yard and whatever he perceives to be the perimeter of our kingdom for signs of a fox intrusion. One whiff of fox urine and this sweet little potato will lose his mind. Sure enough, he finds a such a spot. His brow furrows. He scrunches his wrinkles up extra hard to make himself look as menacing as possible. He looks around indignantly, then angry pees, kicking up grass behind him to make sure that fox knows whose area this *really* belongs to. After a number of angry pees, he's ready to return to home base. I let him inside while I go back out to fetch the paper.[3]

As I pass by our mason bee hotel wedged in the magnolia tree's lower branches, I see a clump of something on the front of the hotel. What is that? A leaf or something? I look a little closer: It's the shriveled-up exoskeleton of a cicada. I scan the tree trunk and spot another. And another. And another. Four. And then I see an actual adult cicada. I stifle a scream—only on account of it being early and the neighbors being asleep—and run back inside to tell Chuck. I'm dancing up and down in my pajamas; he is barely awake and slightly less enthused by my discovery.

This particular cicada must have molted hours ago. It's moving up the tree, already looking for a mate. When they first molt, they look like Nosferatu[4]—white and creepy with red eyes. They hang

---

[3] Yes, I do know that HE is the dog and HE is supposed to be fetching the paper, but fox surveillance and angry peeing must take a lot out of him because after all that, he just wants to go back inside and eat breakfast.

[4] From a silent film in 1922. He was a super bad vampire.

from their tissue-thin husks and wait for their wings to dry and colors to darken. Only then is it safe for them to move about.

So why do they exist? Who knows, really? I'm just spitballing here, but cicadas do serve certain environmental functions. For one, they get eaten. Birds, racoons, opossums, foxes, mice, shrews, frogs, toads, turtles, fish, other insects, and people eat them. Studies have shown that some bird populations grow significantly after a periodical cicada emergence. The bounty is so plentiful, birds can afford to lay more eggs and have a steady food source for their chicks. When cicadas molt, they are especially vulnerable. Birds can come along and pluck them right off tree trunks.[5]

It's interesting. When millions and millions of individuals synchronize their mass emergence, their sheer numbers tend to overwhelm predators. All the predators that eat cicadas get their fill, and there are still plenty of cicadas left over to breed the next generation. This is called predator satiation.

From the moment I see that first cicada, I'm on high alert. No, I don't scrunch up my face and angry pee on anything, but I do check the yard and woods often, looking for cicadas in different stages—some tunneling out of the ground, some molting, others racing each other up trees. Within days, our trees are fairly covered. It isn't scary, though. It's amazing.

It's hard to put this in context for you, just the sheer number of cicadas that emerged. They are literally everywhere—trees, sidewalks, houses. When you walk into your house, you have to do a spot check because the likelihood that you're carrying a hitchhiker

---

[5] Even pets like dogs and cats have been known to indulge. Cicadas aren't toxic to pets, but I was mighty relieved that Zorro was disinterested in them. I didn't relish cleaning up his rainbow cicada poops later.

is extremely high. They are literally choking the skies and clouding up Doppler radar. The National Weather Service issued a statement, "You may have noticed a lot of fuzziness . . . on our radar recently. Our guess? It's probably the cicadas."

You're not just *seeing* them, you're *hearing* them too. Brood X is made up of three species—*Magicicada septendecim*, *M. cassinii*, and *M. septendecula*. The males of each of these species have a special song to call to the females.

The song of *M. septendecim* is otherworldly. It sounds like a flying saucer from a 1950s science fiction film. *M. cassinii*'s song is a little more sassy—it's like throwing water in a vat of hot oil. *M. septendecula* sounds like an angry squirrel to me. Together, their sounds make up a chorus like no other. The din can top 100 decibels. That's louder than a lawn mower. It's a sound you can't quite get away from. We can hear it from *inside* our house.

If a female cicada is interested in any of her many, many suitors, she coolly makes a clicking sound with her wings. She can afford to be choosy. There are more males flitting about than females. After mating, she will lay up to 600 eggs in batches of 20–30. She looks for pencil-thin tree branches, makes a small slit in the bark, and lays her eggs there. In six to eight weeks, the eggs hatch, the nymphs crawl to the ground, dig a hole, and burrow in to do their tree-root-sucking routine for another 17 years. See you in 2038, suckers!

Of course, that's if all goes well. Cicadas have a lot of predators to dodge, but they also have something else. *Massospora cicadina*. That's a fungus. It lies in wait in the soil and infects cicada nymphs as they crawl out of the ground.

Any interaction with this fungus is bad news for the cicada. The fungus will control the cicada's brain while it slowly eats the cicada alive. Within a week of contact, the fungus has buried itself within the cicada's abdomen. What happens next . . . well, I'm just going to say it: The cicada's butt falls off. That lower part of the abdomen just falls right off. In its place, the fungus installs a white, chalky plug.

You'd think the cicada might mind. That it might be a little *unnerved*, at least, that it's lost a third of its body. But no, the cicada carries on as if nothing unusual has happened. That's because the poor little insect is drugged out of its mind. This insidious fungus is packed with an amphetamine called cathinone. It amps up the cicada. When the cicada goes about its business and flies around, fungal spores literally rain down on the ground and onto unsuspecting cicadas. West Virginia University expert Matt Kasson calls the infected cicadas "flying saltshakers of death." The infected cicadas don't seem to experience any pain, but, obviously, this fungus is fatal to them.

You can imagine my absolute horror when I find a fungus butt hanging around in my front yard during week three of the invasion. It was just as I had read about: The whole back end of the cicada had been replaced by a white, dusty plug. He didn't seem to care one way or the other. Only about 5% of the cicadas are estimated to get this fungus, so it doesn't threaten the continuation of the cicadas.

Oddly enough, the cicadas seem to be the only thing this fungus needs to survive. Think about that for a second. An insect we rarely see is the driving life force behind a fungus we've never heard of. Chuck keeps asking me "Why?" I offer this up as a possible explanation. He doesn't seem impressed. "This is all for a *fungus*?"

If animals aren't eating cicadas and a fungus isn't rotting their butts off, then cicadas die a natural death. When they do, their little bodies pile up. That's good news for trees. As the cicadas decompose, they add nutrients to the soil. It's their way of giving back, I suppose. And cicada exit holes aerate the soil beneath trees, which enhances rainwater filtration. So there's that.

FEAR AND LOATHING

All in all, cicadas are weird, but they are kind of helpful as food sources, hosts, and fertilizer. They don't affect people too much. They can't harm us. They don't bite. They don't sting. So why is the news coverage of cicadas always so alarmist? The word *invasion* keeps coming up. Headlines like "They're Coming!" certainly don't help. It seems that some people are afraid of cicadas.

We're all afraid of something. Orthodontists. Ventriloquist dummies. Being marooned on another planet. For some people, it's insects. We know that a lot of people are uneasy around insects. Seeing so many cicadas in such large numbers all at once could be unsettling. For some people, though, the discomfort goes beyond the heebie-jeebies. For some, the fear is a full-blown phobia. A phobia is an extreme fear of a thing, place, or situation.

Being scared is complicated. We have a lot of words for it, and they sometimes indicate our degree of discomfort: *unease, apprehension, misgiving, consternation, angst, trepidation, anxiety, worry, dismay, dread, fear, panic, horror, terror.* How do you know if you have an aversion to something versus an all-out phobia?

The American Psychiatric Association has a gold standard for this. It's outlined in their *Diagnostic and Statistical Manual of Mental Disorders.* You probably don't spend a lot of time with this document. Nor do I. I did look it up, though, and it lists five criteria for a person to qualify as being phobic. The person's experience of fear should be reliable, persist for at least six months, induce significant impairment, and cause either avoidance of the thing or endurance with extreme distress. The final criterion is that the person recognizes their fear to be unreasonable or excessive. That last one is tough because many people who have phobias *do not* think their fears are unreasonable. Their distorted view of the thing they fear produces a heightened sense of impending danger, so *of course* they should be afraid of it.

Are you afraid of belly buttons, garlic, or the pope?[6] No? Some people are. Some people have an extreme fear of long words. It's called hippopotomonstrosesquippedaliophobia.[7] It's very hard for a person to get help for that one because they can't tell anyone what their fear is without triggering a panic attack. How about this one? Nomophobia. That's the fear of being without your cell phone. People with nomophobia experience excessive anxiety about not having their phone with them, their battery being low, or their phone being out of service. I think I know people who have this phobia, and I bet you do too. Ten of the most common phobias are fear of open spaces, heights, flying, enclosed spaces, *insects*, snakes, dogs, storms, and needles.[8]

About 1 person in 10 develops a phobia in the course of their life. Nearly 50 million people experience anxiety involving animals, and 11 million people wrestle with entomophobia. Women seem to develop phobias at twice the rate of men. We think. Men sometimes lie about this kind of stuff—at least the tough guys do.

---

[6] Phobias known as omphalophobia, alliumphobia, and papaphobia

[7] It also happens to be one of the longest words in the dictionary.

[8] Fear of open spaces (agoraphobia), heights (acrophobia), flying (pteromerhanophobia), enclosed spaces (claustrophobia), insects (entomophobia), snakes (ophidiophobia), dogs (cynophobia), storms (astraphobia), and needles (trypanophobia)

Anyone can develop a phobia. The surrealist painter Salvador Dalí was so afraid of grasshoppers, he once jumped out of a window to escape one. Actress Scarlett Johansson is terrified of cockroaches. For actress Nicole Kidman, it's butterflies. The great movie director Steven Spielberg is allegedly phobic of all sorts of insects. If you blanch at the sight of a bug, you're in good company.

For those people with entomophobia, there are a number of triggering insects out there:

Fear of moths: *mottephobia*
Fear of termites: *isopterophobia*
Fear of ants: *myrmecophobia*
Fear of lice: *pediculophobia*
Fear of wasps: *spheksophobia*
Fear of bees: *apiphobia*
Fear of cockroaches: *katsaridaphobia*
Fear of grasshoppers: *orthopterophobia*

What are people afraid of? There are a number of possibilities. We might worry that insects can invade our homes or our bodies. We might be creeped out by their quick, unpredictable movements. We might be freaked out that they can rapidly reproduce and create staggeringly large populations. We might fear them because they can harm us directly (biting or stinging) or indirectly (transmitting disease or destroying our food, electrical wiring, woodwork, carpet, etc.). We might just be disturbed by their ickiness or alienness.

For phobics, the fear is a heart-pounding response to what they perceive as a clear and present danger—that sudden sense of terror or panic that may come with trembling, sweating, dizziness, hyperventilating, or all these things at the same time. Phobics may also experience an overwhelming urge to escape.

I have yet to encounter anyone afraid of cicadas, but Dr. Greta Hirsch certainly has. Dr. Hirsch is the clinical director at the Ross Center in Washington, DC. She specializes in using cognitive-behavioral therapy (CBT) for panic disorder, social anxiety, post-traumatic stress disorder (PTSD), depression, and other anxiety and mood disorders.

During her long career, Dr. Hirsch has treated many patients with phobias—people who fear snakes, insects, flying, vomiting, driving on a highway, crossing a bridge . . . She said she's never had to treat anyone for cicada phobia when it isn't cicada season, but during that last cicada round in 2004, she did have people calling her for help.

According to Dr. Hirsch, there are a number of ways a person can develop a phobia. Perhaps the person lived through a negative incident centered on their phobia, like getting stung by a bee and going into shock. There could be a family connection to the phobia—if a person's parent exhibited fear when they saw a wasp, then the person may have developed a phobia of wasps as a learned response. Or the phobia could have come from the fact that the person grew up in the city and never saw ants but now sees them all the time in the suburbs, and they make the person anxious. The person may even be feeling something innate, a deep disgust of beetles, perhaps, that leads to their fear.

I think about this. I don't have any issues being around cicadas, but *grasshoppers* are another matter. I'm pretty sure it's not a phobia, though, because phobias are *irrational* fears that cause you to alter your behavior. And while I definitely alter my behavior to avoid them, everyone knows that grasshopper fear is *not* irrational. It's totally real. And justified. *Of course.*

That said, I *can* pinpoint a traumatic episode connected to a grasshopper. I was about nine years old and playing outside. I saw a grasshopper on a rock. It was very beautiful, but I was fearful of it because grasshoppers are so unpredictable with all that jumping around.

I wanted this particular grasshopper to go away, so I threw a stick in its general direction. Anyone who knows me will tell you that my hand-eye coordination is dismal. I have the aim of a *Star Wars* stormtrooper—I never hit *anything*. But for some reason, that poor grasshopper took a direct hit. The stick struck the grasshopper and knocked off one of its back legs. The grasshopper leaped away, but the leg remained on the rock, twitching. I was horrified. I felt a deep sense of shame that I had hurt this little animal. From that day on, my fear of grasshoppers was set. I *knew* the other grasshoppers would find me to avenge their comrade eventually.

Every patient is different, which is why Dr. Hirsch learns her patient's history to try to uncover the root of their anxiety. With all her patients, she begins by asking them to keep an anxiety journal. In it, the patient describes any situation that made them feel anxious, ranks their anxiety level, records any physical symptoms they might be feeling, and writes down any thoughts about that. Together Dr. Hirsch and the patient create a fear hierarchy. What are the patient's worst fears? Mine, obviously, was that vigilante grasshoppers would hunt me down and exact their revenge. It would be quite terrible, whatever they decided to do.

For treatment, Dr. Hirsch would begin something called exposure therapy. It's designed to gradually desensitize a person to their fear. One way to measure a person's fear level is to see how near they are willing to come to the feared object. A first step might be to look at a picture of the thing that is feared. For some patients, this can be very difficult. They become anxious right away.

Apparently, not me. I can look at a picture of a grasshopper. It creeps me out, but I can do it. Could I do any of the next steps? Read about grasshoppers? Yes. Learn about their life cycles? Sure. Look at a dead grasshopper in a jar? Probably. Be outside near a live grasshopper? It's happened before. Hold a live grasshopper in my open palm? Uh . . . no. Hard stop on that last one.

"Most anxiety is about a high need for control," Dr. Hirsch says. "A lot of phobias are about anticipatory anxiety." A person might be afraid to go outside because they think their car might be covered in cicadas (or grasshoppers). It's not like they are standing outside seeing that their car is actually covered. It's the fear that their car *might be*. People with phobias deal with a lot of what-ifs.

"Sometimes the anxious mind confuses the possibility of something happening with it being a highly probable event," Dr. Hirsch says. Just because something can happen doesn't mean it probably will. In my case, the possibility that a gang of grasshoppers might find me and overtake me . . . I mean . . . when you say it out loud like that . . .

Dr. Hirsch tells me that many patients with phobias have some aspect in their lives that feels out of control. The phobia develops as a way for the mind to focus on something else instead of the thing that's really bothering them.

I wonder about that. There are a lot of things in my life that I don't feel in control of, but is it possible that I don't actually *fear* grasshoppers? True, they fill me with a sense of dread. I think, perhaps, what I'm feeling isn't fear or anxiety. It's guilt.

Hmmmm . . . something I'll have to ask Dr. Hirsch about.

FEAR AND LOATHING, FOR REAL

Remember, phobias are irrational fears. But that's not to say that there aren't legitimate circumstances when fear and loathing of insects is inappropriate. Insects can, in fact, be a threat to us. They can kill our crops. They can carry diseases. They can even be used as weapons of war.

Let me illustrate my point by telling you a really scary (but true) story. It goes like this:

The year was 1875, and *Melanoplus spretus* had just reached Nebraska. When the locusts came, the sky turned silver. The sun reflected off their wings, making the malevolent horde sparkle. The sounds they made was inescapable—the clicking of their wings, the gnashing of their jaws.

To defend themselves against the invasion, farmers rushed to cover their wells. They blanketed their vegetable gardens with quilts and other bedding. Such foolish notions. The Rocky Mountain locusts easily fouled their water supplies. They chewed through fabric barriers and devoured the plants beneath. Entire fields of crops were leveled in a matter of hours. Nothing was spared—wheat, corn, melons, tobacco, barley, potatoes, beans, fruit trees. It wasn't enough. No. A visit from a swarm meant total loss.

"The air is literally alive with them," a *New York Times* correspondent wrote at the time. "They work as if sent to destroy."

In their frenzy, they inhaled nearly every organic thing. They devoured saddles. They gnawed the handles off farm tools. They gobbled paint off houses. They ate laundry drying on the line. They chewed the wool right off sheep.

The swarm took five days to pass. It was not the first swarm of locusts to decimate the land, nor was it to be the last. But it was the largest recorded locust plague in North American history. Albert Child, a county judge and sometimes meteorologist in Plattsmouth, Nebraska, had done the math. By telegraphing for reports from surrounding towns and timing the rate of movement as the insects streamed by, he estimated the swarm was 2,896 kilometers (1,800 miles) long and 177 kilometers (110 miles) wide. It covered a swath of land greater than the area of California. It may have numbered some 12.5 trillion individual insects.

Between 1873 and 1877, *Melanoplus spretus* caused $200 million in crop damage in swarm after swarm in Colorado, Kansas, Minnesota, Missouri, Nebraska, and other states. During this period, farmers fought back hard. They raised chickens and other birds to eat the locusts. The birds couldn't keep up. *The farmers* ate the locusts, pan-frying them in butter and seasoning them with salt and pepper. But how many locusts can a person reasonably eat? They dug trenches around their fields, filled them with coal tar, and set them on fire. The masses of locusts quickly smothered the flames. Farmers smacked locusts with sticks. They built contraptions to vacuum them. They flooded them. They blasted them with shotguns. They even used dynamite—making locusts probably the only pest ever to be battled with explosives.

Nothing worked. The onslaught continued. They seemed

all-powerful. The locusts could literally stop a train. During the day, railroad tracks would absorb the sun's heat. Locusts would settle on the warm tracks at night. By morning, the tracks would have cooled, and the locusts would be stiff and slow-moving—too slow to move from the path of a speeding train. As early morning trains would hurdle by, the locusts would be crushed on impact. The sheer volume of slick and sticky bodies was enough to reduce a train's traction to stop it altogether. The Rocky Mountain locust made life untenable, rendering the land useless. Business ground to a standstill. Day-to-day activities became impossible.

DR. JEKYLL, MEET MR. HYDE

All this ruination because of an insect? You might be wondering: *What exactly is a locust? Is it just a particularly vicious species of grasshopper?* All locusts are grasshoppers but not all grasshoppers are locusts. I'm not just playing with semantics. The distinction is quite important. Grasshoppers naturally progress through three growth phases—egg, nymph, and adult. However, in *certain* grasshopper species, there is fourth and final phase: the locust. It is during this final phase that grasshoppers alter their shape, size, color, and behavior.

A grasshopper is a solitary thing. It's a bit sluggish, owing to its low metabolic and oxygen-intake rates. It doesn't socialize much. It blends in with its surroundings so it can go unnoticed. When a grasshopper transforms into a locust, a Dr.-Jekyll-and-Mr.-Hyde sort of thing takes place. In a matter of hours, the grasshopper's colors deepen. He's no longer brown and dusty green. He's black with yellow or orange highlights. And that's just the beginning of his ghoulish changes.

The grasshopper's whole body transforms. His shoulders

become broader. His wings become longer. His metabolic and oxygen-intake rates soar. The grasshopper is turning into a totally different grasshopper. It is aggressive and agitated. It doesn't want to be alone. It wants to seek out other like-minded grasshoppers to swarm with. Entomologists have a term for this. They call it the gregarious phase. Seems like a bit of a misnomer to me. We aren't talking about a chatty, life-of-the-party insect. This is more of an insect with murder and mayhem on its mind.

The way solitary grasshoppers and gregarious locusts look and behave is so different, they were thought to be separate species until 1921. Of the 8,000 species of grasshoppers in the world, only about 12 are known to have this capacity to transform, including the Rocky Mountain locust. It took scientists a long time to figure out what drives this change.

Have you ever heard of serotonin? It's a neurotransmitter that allows brain cells and nervous system cells to communicate with each other. In the human body, serotonin does all sorts of interesting things. It helps stabilize our moods, helps us sleep, and helps us digest our food. Would you believe that grasshoppers have serotonin too? They do! A group of scientists in the UK and Australia discovered that when the serotonin levels of certain grasshoppers spike, they turn into Mr. Hydes.[9] But what causes a spike?

In the wild, we know that grasshopper populations explode after rainy periods because there's plenty to eat. When all that food gets eaten, the search for food to sustain the swarm brings grasshoppers together in smaller and smaller areas. Everyone is looking

[9] The "Jekyll and Hyde" versions of the locust are an example of something called phenotypic plasticity. This happens when the genes of the animal go unchanged, but the behavior and physiology change in response to external factors. You probably don't need to know that, but I thought it was pretty interesting.

for a meal, and now everyone is on top of each other. As in many animal species, crowding can trigger changes.

The UK and Australian researchers looked into this. They discovered that the way to send a solitary grasshopper into a murderous frenzy is by tickling its back legs. No, I am *not* making this up! How could anyone make *that* up? C'mon! It's too weird.

Ask me how they figured this out. Go ahead, ask me! You're going to love this. They rounded up 170 desert grasshoppers (*Schistocerca gregaria*). Each grasshopper was placed in a clear plastic cage that was capped on both ends by a wire mesh. The researchers slipped a tiny paintbrush through the mesh and systematically touched different body regions for five seconds every minute for four hours.[10]

Touching the mouth parts, face, antennae, feet, sides, or abdomen had little effect on the grasshoppers. This is because these parts are normally touched by a grasshopper when it is cleaning or eating or even walking around. No, it was those muscular back legs that triggered the flood of serotonin.

If you think about it, this makes sense. When grasshoppers are in their solitary mode, they don't want to be touched. Normally, the hind legs would be used to fend off another grasshopper if it came too close. In a swarm situation, grasshoppers are packed together. The tighter the space, the more jostling occurs.

What happens when jostled grasshoppers get jacked up on serotonin? Instead of being repelled by each other, the grasshoppers are now attracted to each other. They seek each other out and travel together in swarms. They become ravenous. They now have stronger bodies and wings to sustain longer and longer flights. They can travel great distances to find food. Which is what happened

---

[10] How's *that* for a job? All in the name of science, folks.

to those poor Nebraskans back in 1875—they were besieged by ravenous locusts.

The Rocky Mountain locust was native to mountain valleys along both sides of the Continental Divide. When a drought there lead to food shortages, the grasshoppers had no choice but to transform. They had to swarm to survive.

A CLEAR AND PRESENT DANGER

Lest you think that locust swarms only occurred long ago, let me tell you that locusts continue to pose a devastating threat today. In 2020, massive swarms raged in dozens of countries, including Kenya, Ethiopia, Uganda, Somalia, Eritrea, India, Pakistan, Iran, Yemen, Oman, and Saudi Arabia. When swarms affect several countries at once, it is known as a plague.

Hundreds of billions of locusts formed swarms the size of major cities. They laid waste to everything in their path. The Food and Agriculture Organization of the United Nations estimated that a swarm covering 1 square kilometer[11] ate as much food in a day as 35,000 people. And these swarms were traveling about 160 kilometers (100 miles) a day. It was the worst outbreak in Ethiopia in 25 years. In Kenya, it was the worst in seven decades.

In this era of human-driven climate change, experts warn that the threat of locust swarms may grow. This is especially true if the conditions for swarms become more common. For instance, an increase in cyclones in Eastern Africa and the Arabian Peninsula, a weather condition favorable to locusts, could lead to more swarms.

The worst consequences may be borne by people who bear little responsibility for the changing climate, such as small-scale

[11] This might mean about 40 million locusts.

farmers and those living in rural communities. Just as with the North American swarms of the late 1800s, today's farmers watched the locusts devour their crops in the field as well as in storage. They tried the same antiquated techniques to fight them off. One invasion in the village of Mathyakani, Kenya, took its toll on the mental health of its people. For more than a week, children could not go to school. They stayed home to help their parents resist the swarm. The children were enlisted to bang on drums and scream at the pests to try to scare them away. As you can imagine, all this commotion stressed everyone out but had little effect on the locusts.

INSATIABLE THREATS

Plagues of locusts are terrifying. But they aren't the only insects that can seriously harm crops. Herbivorous insects destroy one-fifth of the world's total crop production every year. On any given day, farmers are doing battle with stem borers, root borers, thrips, weevils, and aphids.

Thrips (in the order Thysanoptera) are tiny, slender insects with fringed wings. They are so small, they are usually discovered by the evidence they leave behind. When thrips target host plants, they feed by rasping on leaves and plant tissues to release sap, which they lap up. As they feed, the cells of the plant collapse, forming pits, distortions, and brown patches on the leaves. Later, the leaves have a silver sheen. Your cucumbers are not safe around thrips. Nor are your tomatoes, corn, soybean, or cotton.

The boll weevil (*Anthonomus grandis*) is such a formidable adversary, farmers, lawmakers, and scientists have worked together for years to defeat it. The larvae's ability to feed on and destroy cotton

is unprecedented. Boll weevils entered the US from
Mexico in the late 1800s,
when they were first spotted in Texas. By the 1920s,
they had spread through all
the major cotton-producing
areas in the US. The chief of
the US Department of Agriculture (USDA) once tes-
tified before Congress that

the insect's outbreaks were a wave of evil. To combat them, one-
third of the insecticide used in the US was dedicated to fighting
boll weevils. Other weevils that wreak havoc are the grain or
wheat weevil (*Sitophilus granarius*), which damages stored grain;
the maize weevil (*Sitophilus zeamais*); and the striped bean weevil
(*Alcidodes leucogrammus*).

Then there are aphids, so reviled by farmers that they are called
plant lice. There are more than 1,350 species of aphids on record
in the US and Canada, of which about 80 species are pests of food
crops and ornamental plants. Most get their names from the plants
they attack, such as the green peach aphid (*Myzus persicae*), potato
aphid (*Macrosiphum euphorbiae*), or cabbage aphid (*Brevicoryne brassicae*). These soft-bodied, pear-shaped insects feed on buds, leaves,
flowers, stems, and fruits with piercing-sucking mouthparts.

Each plant reacts differently to aphid attacks. Damage can include
decreased growth rates, mottled leaves, leaf yellowing, stunted
growth, browning, wilting, low yields, and death. If that's not
bad enough, aphids are prolific poopers that churn out something

euphemistically called honeydew. This sticky substance drips onto plant leaves and stems and can harbor fungal diseases like powdery mildew and black sooty mold. It's all a little unpleasant.

Having aphids attack your crops is only half of your problem, though. Due to the way they feed, aphids can transmit bacterial and viral diseases. The green peach aphid, for example, is a vector for more than 110 plant viruses. Aphids may transmit viruses from plant to plant on squash, cucumber, pumpkin, melon, bean, potato, lettuce, beet, chard, and bok choy.

Farmers know they lose crops to pests and plant diseases, but scientists have found that on a global scale, crop yields for the five major food crops are reduced by 10–40%. Wheat, rice, maize, soybean, and potato yields account for about 50% of the global human calorie intake. Food security for the entire planet is at risk, which is kind of strange to think about when you know that we'd all be dead without other insects pollinating our crops.

AS DISEASE CARRIERS

We know that insects can be destructive to plants. They can also be vectors, which means they can carry diseases. Mosquitoes, flies, midges, bed bugs, lice, and fleas can all be vectors. Each are capable of transmitting a number of sicknesses caused by viruses, bacteria, and parasites. I'm afraid if I start talking about this, you'll never want to leave your home again because you'll be worried you'll contract something awful. It's a risk I'm going to have to take because you do need to know the harm that insects can do to us.

We'll start with the worst offender: the mosquito. Everything you might need to know about this beast can be found in *The Mosquito: A Human History of Our Deadliest Predator* by Timothy

C. Winegard. Wingard's nearly 500-page tome tells us that the mosquito has killed more people than any other cause of death in human history. This tiny insect has dispatched an estimated 52 billion people from a total of 108 billion throughout our 200,000-year existence. It's not the mosquito, per se, so much as the diseases she carries. And yes, we can say "she" here because male mosquitoes don't bite. Only the females do and only right before they are ready to lay eggs.

There are more than 100 trillion mosquitoes in the world at any given moment. Of the approximately 3,500 species, only a couple hundred feast on human blood. Those that do can deliver some very nasty diseases.

Cribbing from the World Health Organization, I tried to make a little chart to keep track of the worst of it.

VECTOR		DISEASE CAUSED	TYPE OF PATHOGEN
Mosquito	*Aedes*	Chikungunya	Virus
		Dengue	Virus
		Lymphatic filariasis (Elephantiasis)	Parasite
		Yellow fever	Virus
		Zika	Virus
	Anopheles	Lymphatic filariasis (Elephantiasis)	Parasite
		Malaria	Parasite
	Culex	Japanese encephalitis	Virus
		Lymphatic filariasis (Elephantiasis)	Parasite
		West Nile fever	Virus

I'm guessing you've heard of some of these before. Of all the mosquito-borne diseases, malaria is the most widespread and the deadliest. It is not caused by a virus or bacteria. It's caused by a parasite belonging to the genus *Plasmodium*. Five species cause the vast majority of human illnesses, but *Plasmodium falciparum* causes the most deaths.

Malaria is not something you want to contract. Symptoms for a *mild* case can include fever, moderate-to-severe shaking, chills, profuse sweating, headache, nausea, vomiting, diarrhea, and anemia. If your case is severe, it can kill you within 24 hours. Children under the age of five are the most vulnerable. In 2019, they accounted for 67% of all malaria deaths worldwide. Malaria can be treated with prescription drugs to kill the parasite.

The *Aedes* mosquito traffics in other terrible illnesses like chikungunya, dengue, elephantiasis, yellow fever, and Zika. Each of these is driven by a virus. Chikungunya brings on fever and severe joint and muscle pain. It doesn't usually kill you—just makes you beyond miserable. The name *chikungunya* is derived from a word in the Kimakonde language.[12] It means "to become contorted" and describes how sufferers look while stooped over with joint pain. There is no vaccine, and the recovery period is long.

Dengue fever often goes by another name—breakbone fever, because of the severe muscle, joint, and bone pains it causes. Dengue is common in more than 100 countries around the world. Forty

---

[12] Spoken by the Makonde in Tanzania

percent of the world's population—about three billion people—live in areas with a risk of dengue. It can be confused with other illnesses that can cause fever, aches, or rash. Most people who get it will recover after about a week, but the danger is that once you've had it, you're more vulnerable to other more serious illnesses. A vaccine is under study.

Elephantiasis is less common but one you've probably seen pictures of. It's kind of disturbing. It starts when mosquitoes infected with a type of roundworm larvae bite you. The tiny larvae survive in your bloodstream and grow. They end up in your lymph system and cause extreme swelling in your limbs. This condition can be treated with anti-parasitic drugs, but it takes a long time to recover.

Yellow fever takes three to six days to develop and brings on fever, chills, headache, backache, and muscle aches. The name reflects the color you might turn if you get it. Jaundice—the yellowing of the skin or whites of the eyes—is one symptom. About 15% of people who get yellow fever can develop internal bleeding and go into shock or organ failure. However, a single dose of yellow fever vaccine is enough to grant you a lifetime immunity against the disease.

In early 2015, a widespread epidemic of Zika fever spread from Brazil to other parts of South America and into North America. Many people infected with Zika virus won't have symptoms or will only have mild symptoms that might include fever, rash, headache, joint pain, muscle aches, and red eyes. What attracted a lot of attention to Zika, however, is that it can be passed from a pregnant woman to her baby. Infection during pregnancy can cause certain birth defects, namely microcephaly. Babies with microcephaly often have smaller brains that might not have developed properly. There is no treatment available for Zika virus.

The *Culex* mosquito can transmit Japanese encephalitis. This disease is now extremely rare and can be prevented by a vaccine. Most people who get infected experience mild or no symptoms, but in extreme cases, it can cause brain swelling and seizures.

And finally, the West Nile virus. West Nile is the leading cause of mosquito-borne disease in the continental United States. Most people do not feel any symptoms, but those who do could experience a whole host of unpleasantness: high fever, headache, neck stiffness, stupor, disorientation, coma, tremors, convulsions, muscle weakness, vision loss, numbness, and paralysis. Older people appear to be at a higher risk of severe symptoms. Vaccine trials are underway.

As you can see, mosquitoes suck.[13] Sadly, they are not the only insects that transmit diseases. Sorry. There are more. Tsetse flies (*Glossina* genus) deliver African trypanosomiasis, better known as sleeping sickness. This one is especially terrifying. The disease attacks the brain, leaving some victims in a statue-like condition, speechless and motionless. It has a high mortality rate not only among people but also among cattle. It can be treated with an antiparasitic drug.

Another awful illness is American trypanosomiasis—also known as Chagas disease. This virus invades the muscle cells of the digestive tract and heart. Infection is most commonly spread through contact with the poop of an infected triatomine bug (or kissing bug), a blood-sucking insect that feeds on people and animals. As people typically show no symptoms for many years, most

---

[13] A gentle reminder from Chapter Two: They do serve as pollinators.

are unaware they have Chagas. Up to one-third of people with Chagas will suffer heart damage that becomes evident only many years later. Chagas kills more people in Latin America each year than any other parasitic disease, including malaria.

It's almost overwhelming to think that these terrible diseases are moving in and around our world every day, relying on insects to do the dirty work of infecting people. Unfortunately, insects get stuck in this role quite a lot. Consider this: Diseases spread by insects have affected armies, influenced the outcome of wars, and determined the fate of empires. Did Rome fall because of insects? Maybe not, but malaria was an albatross around the neck of Roman civilization.

And though early historians tried to blame rats, we know that fleas were the source of the Black Death. It was a devastating global epidemic of bubonic plague that struck Europe and Asia in the mid-1300s. In 1347, 12 ships from the Black Sea docked at the Sicilian port of Messina. Most of the crew were dead and those still alive were mortally ill. They were covered in black boils that oozed blood and pus. Little was known at the time, but we now know that the bubonic plague is spread by a bacterium called *Yersinia pestis*. It's transmitted through the bite of an infected flea. Over the next five years, the Black Death killed more than 20 million people in Europe—almost one-third of the continent's population.

The main insect-borne diseases of importance in early times were plague, typhus, yellow fever, and malaria. Without a doubt, these illnesses altered the course of history. Few dispute that Napoleon's campaign against Russia in 1812 failed because of outbreaks of louse-borne typhus or that his early operations in the

New World had to be aborted because of yellow fever and malaria. Look at the numbers from the Crimean War (1853–56). There were 730,000 British, French, and Russian combatants. Of those, 34,000 were killed in action; 26,000 died from wounds; and 130,000 died from diseases, such as cholera and typhus. Typhus is contracted from infected mites, fleas, or lice. The long stalemate between the German and British troops in the Gallipoli campaign (1915–16) in World War I was due to the large number of men in each army who were incapacitated by malaria. Two-thirds of the approximately 660,000 deaths of soldiers from the American Civil War were caused by uncontrolled infectious diseases—many brought on by insects—such as typhoid and malaria.

World War I soldiers could not escape trench fever, transmitted through body lice. One study in 1915 found that 95% of soldiers were infested with an average lousiness of 20 lice per man, with 5% having 100–300 lice each. Lice primarily infest clothes. The trenches kept men huddled close together, so the lice infestations spread and spread. World War II was probably one of the first wars in which diseases like typhus and malaria killed fewer people than bullets and bombs—and that was only because everyone was sprayed down with DDT. Insects, surely, have been the cause of many deaths.

WAGING WAR WITH INSECTS

What about the times when insects have been used in war *on purpose* to inflict damage on the enemy? There are three types of war tactics involving insects. The first type uses insects, such as wasps or bees, for a direct attack on opponents.

The Bible's Old Testament, for instance, says that God used

insects in this way. The Book of Exodus recounts how Moses was supposed to lead the Jewish people out of Egypt for Israel. They had been enslaved by the pharaoh, and he did not want to lose his free workforce. Moses asked, but the pharaoh said no.

So the Lord sent down 10 plagues until the pharaoh changed his mind.

What plague was sent first? Blood. Moses went down to the Nile River, whacked the water with a stick, and turned all the water to blood.

Scientists love to try to figure out how something like this might have happened. In this case, their best guess is that the "bloody water" might actually have been caused by a red algae bloom. Under the right conditions, a type of microscopic algae can reproduce so rapidly, the water turns red.

Exodus 7:18[14] tells us that "the fish in the Nile shall die, and the Nile will stink, and the Egyptians will grow weary of drinking water from the Nile." Which pretty much tracks with the scientists because algae blooms contain a toxin that can build up in shellfish and poison the animals that feed on them.

What next? Well, the Lord sent frogs. The scientific explanation would suggest that if the Nile is choked with algae, the frogs would leave and look for a better place to live.

The absence of frogs would then trigger plague number three: gnats—in the form of biting midges. Midges breed in damp, nutrient-rich soil. With their main predator gone, the midge population would probably explode.

The fourth plague is flies. Conditions would certainly be right for them. Flies feed on rotting organic matter, which you'd have aplenty with all the decaying vegetation, fish, and frogs.

[14] English Standard Version

The fifth and sixth plagues—livestock diseases and boils—could have been spread by the midges and flies. Flies can carry anthrax, which would cause the boils.

God sent hail next, which would spark a lot of plant growth. Guess which insect would thrive under those circumstances? That's right: locusts, the eighth plague. Swarms of locusts can be so extreme they can blot out the sun, bringing about the ninth plague, darkness.

Which leaves us with the final plague. God said if the pharaoh did not release his people, he would kill all the firstborns in the land. Frankly, I would have caved a long time ago. At the first sign of a bloody river and frogs popping up everywhere, I would have set those people free. But the pharaoh had a hard heart and would not be moved. So God followed through.

This one is a bit harder for scientists to explain, but one theory centers on grain. The Egyptian grain storage system was vulnerable to moisture and insect infestation. If the grain got contaminated, and the first scoop of grain always went to the firstborn . . .

Do we know if any of this happened in this way? Not for sure. But it's interesting to think about. What we do know for sure is that people have used insects as weapons since ancient times.

BEE BOMBS AND OTHER WARFARE

The master of all things war and insect is none other than the great Jeffrey A. Lockwood. Lockwood is an author, entomologist, and University of Wyoming professor of natural sciences and humanities. His *Six-Legged Soldiers: Using Insects as Weapons of War* tells us that insects have long been used as weapons of war.

Did you know that a bee once started a war? Oh yes. Back

in the day. It was June 24, 637 AD, to be precise—the Battle of Moira (or Magh Rath) in Ireland.

The trouble began when Congal Cláen, king of Ulaid, was visiting his foster dad, Domhnall, High King of Ireland. While walking along the estate, Congal was stung by one of Domhnall's bees in the eye. Ouch! Worse, the bee blinded Congal. Congal then got tagged with the unfortunate nickname Caech, which means "one-eye." Not very sensitive, but people were less so back then.

Congal's family took offense, and they felt that one of Domhnall's sons should be blinded to even the score. Domhnall wouldn't agree to that. Eventually, things got pretty heated. Domhnall offered to kill all his bees as an appeasement, but that gesture did not appease anyone. It was decided that the dispute would best be settled on the battlefield. And some 50,000 men reportedly turned up to fight for one faction or the other. They fought for six days, but the battle resulted in a decisive victory for the High King and his army. Congal (and his singular eye) was killed in the fighting.

It's rare that an insect starts a war, but it isn't rare that an insect is used in war, especially when we are talking about lobbing things at the enemy. Men in war love to throw things at each other. Like bees.

We know that back in the Stone Age, early man resorted to hurling beehives into each other's caves when there was a dispute. While that's not very neighborly, that was only the beginning of the range of human inventiveness when it comes to co-opting insects for war.

By 2600 BC, the Maya had created a clever way to weaponize bees. One of their sacred texts, the *Popol Vuh*, outlines an ingenious

booby trap to ward off invaders. Dummy warriors were decked out with cloaks, spears, and shields and posted along the walls of citadels. War bonnets were placed on their heads, which were actually hollowed-out gourds stuffed with bees or hornets. When the attackers smashed the gourds, the enraged insects gave them what-for. This created just enough chaos for the real Maya warriors to dispatch their attackers.

That's a pretty good tactic if you're waiting for the enemy to come to you. But what about if you are on the battlefield and need to deploy your weapon in the moment? At about the same time the Maya were fashioning their bee-headed warriors, people in the Middle East were molding special containers from clay that were light enough to throw and fragile enough to shatter on impact. You know where this is going, right? Bee grenades! Pack some bees in a pot, carry it into combat, and toss it at your enemies accordingly. Think of it as Operation Fling and Sting. Handy, right?

Well, yes and no. The reality is, stinging insects make for pretty unreliable combatants. They don't exactly follow orders. Nor do they understand who is the Good Guy and who is the Bad Guy during a particular engagement. If you didn't handle your bee grenades properly, you might find yourself on the receiving end of their wrath. And if you got too close to the enemy when you lobbed one, you could get caught up in your own attack.

Better to put some distance between you and your bee bomb. The Tiv people of Nigeria had a more refined technique. They stored their bee ammunition in special horns that also held a poisonous powder to make the bees even more lethal. Then the Tiv would aim the horns at their enemies and blow. As long as they remembered to blow out and not suck in, the Tiv could effectively launch the bees at their opponents from a safe distance.

But pew-pewing a handful of bees at a time can't possibly cause *that much* damage to the enemy. How to create more havoc with more bees? The Romans locked on to an enhanced delivery system early on: They *catapulted* bees. They launched massive insect payloads of bagged or basketed bees. So widespread was the Roman use of catapulted bee bombs that you can actually track the decline in the number of hives during the latter part of the Roman Empire—a consequence, no doubt, of having heaved too many hives.[15] This method was even adapted for use on the high seas. It became one of the most effective ways to clear an enemy ship of topside sailors.

Bee bombs did create something of a supply and demand issue, though. How to get access to fresh bees when you needed them? Beginning in the 12th century, many Europeans solved this problem. Small alcoves known as bee boles were installed in Medieval buildings across Europe. During peace, it was lovely to have pollinators closer to gardens and access to fresh honey. During a siege, however, those bee boles were raided. The hives would be thrown over the walls to terrify and torment attackers. Bombus away!

---

[15] Other delivery systems included a sort of windmill-like contraption that shot straw beehives from the ends of rapidly spinning arms, similar to how a Gatling gun propels bullets.

I don't know how well you remember your world history, but there have been a number of skirmishes where the outcome was determined by bees. When the residents of Chester, England, were assailed by Danes and Norwegians in the early 900s, they were confident that their attackers could not penetrate the city's fortifications, which included a sturdy wall. No one could have predicted that the determined Scandinavians would start tunneling under it. With the ferocity of over-caffeinated honey badgers, they ruthlessly tore away at the earth to breach the town within. Out of sheer desperation, the townspeople collected all of the city's beehives and tossed them into the tunnel. This unexpected move drove off the oppressors.

The lesson learned there was not fully retained by the Scandinavian army, however, as they were once again repelled by bees while storming the walled city of Kissingen (Germany) during the Thirty Years' War (1618–48). It's true that this time, the soldiers were well-protected by heavy clothing and armor, but their horses weren't. When the townspeople chucked beehives over the wall, the Scandinavian horses lost their minds. The attack collapsed, and the city was spared.

Bees have been co-opted many times throughout history. They even came to the defense of some defenseless nuns in the town of Wuppertal, Germany, in the early 1600s. When the town came under attack by marauding troops, the sisters refused to open the gates. The soldiers attempted to enter by force. Not a good move on their part. The feisty nuns toppled all the beehives in their apiary and then ran indoors. It did not take long for the soldiers to be driven back by the masses of displeased bees. Afterward, the town changed its name to Beyenburg (or bee-town) in honor of their defenders.

Lest you think bee battling is a thing of the distant past, the Vietcong also conscripted the Asian giant honey bee (*Apis dorsata*) during the Vietnam War (1955–75). Soldiers painstakingly relocated bee colonies to trails used by American soldiers. They set a small explosive charge near the bees and waited for the Americans to arrive. When an enemy patrol passed by, a Vietcong soldier set off the blast, creating a bee blitz.

Not to be outdone, the US military funded a secret research project during this war to build an apparatus that would spray Vietnamese soldiers with the alarm pheromone of bees. The hope was that this chemical signal would incite bees to attack. They never quite got the kinks worked out on this one, so this weapon was never deployed—but it wasn't for lack of trying.

I don't want to give you the impression that bees have been the only insect used for violent purposes. There was that whole unpleasant business with Nasrullah Khan, the 19th-century emir of Bukhara (present-day Uzbekistan). The emir was part of the Great Game in which England and Russia vied for control of Central Asia in the 1800s. Khan was a cruel and sadistic man, nicknamed the Butcher by his own people.

Colonel Charles Stoddart arrived in 1838, sent to try to arrange an alliance between Khan and the British East India Company against the Russian Empire. However, Stoddart was not schooled in the local customs of such a place and made many faux pas. First, he rode into the castle on horseback rather than walking. *Not cool*, Khan thought. It was customary for visiting dignitaries to dismount, leave their horses outside, and bow before the emir. Stoddart also arrived without a gift for the emir. Unforgivable! The emir promptly imprisoned Stoddart. Not locked up in a cell at Zindon Prison, mind you. No, the emir had Stoddart thrown into his bug pit.

This dark well was 6 meters (21 feet) deep. It was covered with an iron grate and accessible only by a rope. And it was filled with assassin bugs. Assassin bugs are 2.5-centimeter (1-inch) carnivorous insects with curved beaks used for piercing their prey. Being bitten is like being stabbed with a hot needle. The digestive enzymes they inject to liquefy the tissues of their prey cause festering wounds.

Months went by before help arrived. Captain Arthur Conolly of the cavalry came alone to try to rescue Stoddart. When Khan realized, however, that Conolly didn't bring a reply from Queen Victoria herself, he was incensed. Conolly soon joined Stoddard in the pit. I wish this story had a happy ending, but, alas, both men suffered terribly, being gnawed on by assassin bugs until the emir had them hauled out of the pit and shot.

CROP KILLERS

When we aren't lobbing insects at each other or having them devour our enemies, we might use them to destroy our enemies' crops. Awww . . . would we do *that*? Yes, I'm afraid we would, as terrible as we are. Starving one's enemy or crippling their economy by unleashing insects to destroy crops comes right out of the villain's playbook.

It'll come as no surprise to you that World War II Nazis were really keen on the idea of decimating crops to starve their enemies. By the summer of 1944, Germany landed on the Colorado potato beetle (*Leptinotarsa decemlineata*) to be its secret weapon.[16] The potato beetle was a notorious, hard-to-kill plant pest.

---

[16] We expect the Nazis to be bad, but the truth is, both the French and the Americans also looked into weaponizing this insect.

The East Coast of England, thought to be the site of some 400,000 hectares of potato fields, was deemed a suitable target. It was estimated that 20–40 million beetles would be needed for full coverage. To meet this need, German resources were diverted to large-scale breeding programs. A number of tests were performed on German soil, but thankfully the program failed, and the beetles never reached English soil in large numbers.

Other countries also tried approaches to destroy the enemy's crops. During the Cuban Missile Crisis in October 1962, leaders of the US and the Soviet Union engaged in a 13-day political and military standoff over the installation of nuclear-armed Soviet missiles in Cuba. What on Earth could that possibly have to do with insects? I'll tell you. During this crisis, the American government was looking for ways to destabilize Cuba and, at one point, considered unleashing an army of plant hoppers. That's right. Cuba had nukes, and we had tiny bugs the size of rice grains.

The goal was to damage Cuba by destroying its most important economic asset—sugarcane. The sugarcane leafhopper (*Pyrilla perpusilla*) transmits a virus that stunts plant growth and causes plant deformities. The Americans also considered sending infected plant hoppers that would destroy Cuba's rice crop. Disaster was avoided when the Soviets removed their missiles and the US agreed not to invade Cuba.

The ability of insects to wreak havoc is well-known, and in the wrong hands, that power can become a dangerous tool. In 1989, a sudden invasion of Mediterranean fruit flies (*Ceratitis capitata*) appeared in California. Crops began to fail. The whole thing seemed rather odd until a covert group of environmental radicals told authorities they were breeding and releasing the medflies— voracious pests of valuable fruit crops—on purpose. They were protesting the use of insecticides on California crops. The infestation was eventually suppressed. But had the group succeeded in establishing this pest, losses could have reached more than $13 billion.

THE STUFF OF NIGHTMARES

There's one other way people can use insects as weapons, and that involves purposefully infecting insects with a pathogen, then releasing them over enemy territory. The goal would be to have the insects bite people and animals to make them sick.

I need to take a pause for a second here. What I'm about to describe to you is not only the stuff of nightmares, but most of us would agree that it is unethical and immoral. Unfortunately, though, it is absolutely true that, throughout history, people all over the world have exploited insects in terrible ways to do terrible things.

For instance, during World War II, Japan used plague-infected fleas to experiment on prisoners. Then the lethal plague fleas were packed into special bombs and dropped over Chinese territories. These actions created an epidemic that took the lives of about 500,000 people. In addition to the fleas, the Japanese also used flies infected with cholera against the Chinese population. The goal was to kill as many people as possible. The Japanese military then developed plans to spread plague-carrying fleas over San Diego in 1945, but the plan was disrupted by the end of the war.

Around the same time as the plague-flea experimentation, German SS commander Heinrich Himmler ordered scientists at the Dachau concentration camp to develop biological weapons using insects. The goal was to weaponize malaria-infected mosquitoes against Allied troops. Scientists also conducted experiments on prisoners using diseases transmitted by lice, house flies, and fleas.

We might have expected these sorts of programs from some of the Axis powers in World War II, but you need to know that even the US conducted significant research into weaponizing insects. In 1954, Operation Big Itch took place at Dugway Proving Ground in Utah. The test was designed to determine coverage patterns and survivability of the tropical rat flea (*Xenopsylla cheopis*) for use in biological warfare as a disease vector. Big Itch proved "successful" because the tests showed that not only could the fleas survive the fall from an airplane but they also soon attached themselves to hosts.

The goal of Operation Big Buzz in 1955 was to determine the feasibility of producing, storing, loading into munitions, and dispersing the yellow fever mosquito (*Aedes aegypti*) from aircraft. Some 330,000 uninfected mosquitoes were dropped from aircraft in E14 bombs over a residential area of Savannah, Georgia. Scientists estimated how many mosquitoes entered houses and bit people.

In 1956, the US government continued its research with Operation Drop Kick, releasing 600,000 uninfected mosquitoes from a plane at Avon Park Air Force Range, Florida. Within a day, the mosquitoes spread a distance of up to 3.2 kilometers (2 miles) and bit many people. In the predominantly Black community of Avon Park, dozens of Black people became ill, and eight people died. In 1958, further tests discovered that mosquitoes could quickly be dropped from helicopters, would spread more than 1.6 kilometers (1 mile) in each direction, and would enter all types of buildings.

You can start to understand how complicated our relationship with insects is. Collectively, their power can be harvested for evil gains. Yet we rely on them for so many things. On their own, they mostly do tremendous good for our planet. Could we live without them? I don't think so, but let's explore that idea in our next chapter.

7.

E IS FOR EXTINCTION

I'M GOING TO tell you something that completely freaked me out. I hope you can stay calm because we can't both be in a panic. You ready? Here it is:

Almost all species that have ever existed on Earth are extinct.

I'm not kidding. It's something like more than 99% of all organisms that ever lived on our planet are no longer here. We're talking maybe four billion species. Gone. Poof!

What do you MEAN that most everything is already dead? How is that possible?

It's complicated. Let's start with Earth. Earth is estimated to be 4.5 billion years old plus or minus 10 years. Ha ha. No, I'm joking. It's plus or minus about 50 million years. When we get into periods of time that stretch into the millions and billions, it is difficult to process.

To the best of our understanding, as the solar system was forming, gravity pulled a bunch of swirling gas and dust into a little ball and that became Earth. As you know, Earth is a dynamic place. It's always changing. Our planet's moving plates slowly reconfigure oceans and continents. Volcanic activity and earthquakes alter landscapes. Wind, water, and ice erode and shape the land. There's a lot going on, and there always has been.

We say that life on our planet began at least 3.5 billion years ago, and we say that because that's the age of the oldest rocks with fossil evidence in them. Our earliest lifeforms weren't much to write home about—microorganisms . . . bacteria and such. But life on Earth became more complex. Over time some things survived, some things evolved, and some things died. Extinction, apparently, is a normal thing. It's a natural part of the evolutionary process. It allows for species turnover. When new species are formed through natural selection and various external forces, old ones go extinct due to the extra competition, habitat changes, and so forth. Basically, the *species* dies of old age. I guess I never really thought about it like that before.

Extinction takes place constantly. At any point in time, animal and plant species become extinct somewhere in the world, and this is roughly balanced out by the evolution of new species. This low-level everyday extinction is called background extinction. Scientists estimate how long, on average, a species lasts from its origination to its extinction. Mammals, for example, have an average *species* life span of about a million years. Given that, we might expect one mammal to go extinct due to natural causes every 200 years or so.

That's if everything is normal, but extinctions can happen on a massive scale if environmental changes occur faster than animals and plants can adapt to them. Mass extinctions are rare events. There have been five documented cases on Earth. Each varied in size and cause, but all of them killed off an overwhelming majority of species living at the time. In the aftermath of such events, there is usually a period of rapid speciation among the things that do survive. They sort of spread out and take over because there is less competition for food, resources, and shelter. Take a look at the five mass extinctions and how they compare:

ORDOVICIAN MASS EXTINCTION

When: About 440 million years ago

Scope: Up to 85% of all living species eliminated

Suspected causes: Continental shift and climate change

What happened: Our first extinction event took place in two waves, roughly a million years apart. To us, it didn't happen quickly, but in geological terms, yes, this was speedy. The first wave was an ice age triggered by the rise of North America's Appalachian Mountains. The whole planet got frosty and sea levels dropped. Many species couldn't adapt fast enough to survive the cold. The second wave hit when the ice abruptly melted. Ocean levels started to rise again, causing marine oxygen levels to plummet, so most everybody in the water suffocated.

There weren't a lot of land animals at the time, so this was largely a water extinction. Small living things such as corals, shelled bra-chiopods, eel-like creatures called conodonts, and trilobites were hardest hit. A lot of things went extinct. The graptolites didn't make it—tiny animals that lived together in groups and shared the same skeleton, which was like an apartment building. Each animal built its own apartment, and these were stuck together to make a colony. There wasn't enough oxygen in the water to feed the phytoplankton, which meant the graptolites lost their food source. Have you ever even heard of a graptolite before? I certainly hadn't. Seems so odd that there were all these things on Earth, but we only know about them because we've found their fossils. Oddly enough, sponges thrived during this extinction. I guess it makes sense. Sponges can tolerate low oxygen levels and temperature changes, and their food sources increased dramatically, what with all the dying going on around them.

DEVONIAN MASS EXTINCTION

When: About 375 million years ago

Scope: Nearly 80% of all living species eliminated

Suspected causes: Cooler temperatures and lack of oxygen in the oceans

What happened: Earth was experiencing more change. Large trees had been evolving and creating Earth's first forests. Yay! Except, the more these plants grew, the more carbon dioxide was sucked from the air. Carbon dioxide is one of those greenhouse gases that trap the sun's heat. Without it in circulation, our planet began to cool. Glaciers formed, trapping water on land, which lowered sea levels and made life in the ocean rather uncomfortable.

To make matters worse, those deep plant roots were breaking up rocks and helping to create soil. Soil! Great! Yes and no. Plant nutrients and minerals got washed into rivers and oceans, which fed microscopic algae. As the algae multiplied and bloomed, they choked the oceans. Low oxygen levels helped create vast dead zones in the water. If you were a brachiopod, trilobite, or reef-building organism, life became untenable. What survived this period were tetrapods. These were four-limbed animals that were transitioning from sea to land. They would eventually evolve into reptiles, amphibians, and mammals.

PERMIAN MASS EXTINCTION

When: About 250 million years ago

Scope: An estimated 96% of all living species eliminated

Suspected causes: Volcanic activity and climate change

What happened: Known as the Great Dying, this extinction was the single worst event life on Earth has ever experienced. Taking place over a period of about 60,000 years, 96% of all marine species and nearly three-fourths of land species died out. Of the five mass extinctions, this is the only one that wiped out large numbers of insect species.

The event's single largest cause was an immense volcanic eruption across Siberia. The total volume released over land and beneath the surface was enough to cover an area the size of the US in kilometer-deep magma. Everything was on fire. As you can imagine, this really messed with Earth's carbon cycle, bringing abnormally high air and sea temperatures—a global increase by as much as 10°C (18°F). Not to mention a more acidic ocean; all that carbon dioxide released by the eruptions was absorbed by the seas. This wiped out oxygen, leaving parts of the seafloor completely oxygen-free. Earth was in a bit of a reset after this, and it took a long time for life to recover. But the stage was set for a group of reptiles called the archosaurs to take over—the forerunners of birds, crocodilians, pterosaurs, and the non-avian dinosaurs.

TRIASSIC-JURASSIC MASS EXTINCTION

When: About 200 million years ago

Scope: More than half of all living species eliminated

Suspected causes: Volcanic activity, global climate change, and changing pH and sea levels of the oceans

What happened: Sigh. More changes for our poor Earth. Up to this point, there had only been one landmass: Pangaea. All the modern continents were fused together into this one supercontinent. At the end of the Triassic period, it began to break apart. As

North America separated from Africa, massive volcanic eruptions in a hot spot at the center of what would eventually be the Atlantic Ocean started belching out volumes of carbon dioxide, effectively quadrupling the levels in the atmosphere. This was almost a repeat of the previous extinction event. The conodonts didn't make it this time, though. These eel-like fishies had impressive teeth, which started showing up in the fossil record, baffling scientists because they mostly found the teeth but not the animal they were attached to. So many things had died in the last extinction that Earth was taking a hard reset, especially with plants. This was the period in which land animals, especially our friends the dinosaurs, were thriving.

CRETACEOUS-TERTIARY MASS EXTINCTION (OR K-T EXTINCTION[1])

When: About 65 million years ago

Scope: Nearly 75% of all living species eliminated

Suspected causes: Asteroid impact and volcanic activity

What happened: This event is the only extinction definitively connected to a major asteroid impact. An asteroid roughly 12 kilometers (7.5 miles) across slammed into the waters off what is now Mexico's Yucatán Peninsula at 72,420 kilometers (45,000 miles) an hour. The impact—which left a crater more than 193 kilometers (120 miles) wide—flung volumes of dust, debris, and sulfur into the atmosphere, singeing all land within 1,448 kilometers (900 miles) and triggering a huge tsunami. The resulting

---

[1] Scientists refer to this as the K-T extinction because it happened at the end of the Cretaceous period and the beginning of the Tertiary period. Geologists use K as a shorthand for Cretaceous because they use the C as a shorthand for an earlier period (the Cambrian). Thought I should clear that up.

dust cloud encircled the planet, creating an impact winter by greatly reducing the amount of sunlight reaching Earth's surface and preventing photosynthesis by plants on land and plankton in the oceans. As plants and plankton died, extinctions moved up the food chain, eliminating herbivores and carnivores alike. That's the part you know. The part that everyone talks about. But there was more to this extinction than that. Volcanic eruptions at the Deccan Traps in India made things worse. Bye-bye dinosaurs. All non-avian dinosaurs went extinct. In their absence, mammals started to fill the gaps, as did the sole surviving branch of dinosaurs: birds.

Which brings us to today and the sixth extinction. I know. I told you there were five. There were five, but we happen to be right in the middle of the sixth. No asteroids so far. Just people. People collectively killing our planet. This is hard to write about because you just want to scream, *HOW DID WE GET HERE?*

A IS FOR ANTHROPOCENE

Let me try to stay calm while I tell you. The unofficial name for where we are now is the Anthropocene[2]—that's the unit of geologic time used to describe when human activity started to have a significant impact on the planet's climate and ecosystems.

When, exactly, did that start? It is hard to pinpoint. Was it

---

[2] The term hasn't been formally adopted by the International Union of Geological Sciences (IUGS), the international organization that names and defines epochs. What the IUGS is looking for is a golden spike—a marker in the fossil record that could mark the Anthropocene. This marker will have to be so significant that it would be detectable in rock layers thousands and even millions of years into the future.

Britain's Industrial Revolution in the 18th century, which created the world's first fossil-fuel economy? Burning fossil fuels enabled large-scale production and drove the growth of mines, factories, and mills. The rest of the world followed suit, eating up as much coal as possible and belching out carbon dioxide at an alarming rate. That might've been it.

Did the Anthropocene begin on July 16, 1945, with the Trinity test in New Mexico—when mankind detonated the first nuclear device and created an atomic mushroom cloud 12.1 kilometers (7.5 miles) high? Was it after we did more testing and obliterated Hiroshima and Nagasaki with nuclear weapons? Surely that left a permanent scar on our planet.

Or was it even later than that, after we started choking our planet with plastic? Will future scientists see all our plastic in the rock strata?

Maybe the Anthropocene was just brought on by the collective weight of 7.6 billion people living on the planet, all needing the same resources.

It's hard to say, but here's what we can agree on. Agriculture, urbanization, deforestation, and pollution have caused extraordinary changes on Earth: carbon dioxide emissions, ocean acidification, habitat destruction, overhunting, overfishing, the introduction of invasive species, and widescale natural resource extraction. Geologically speaking, humans have been a blip on the planet's radar. Yet we have fundamentally altered the physical, chemical, and biological systems of our world—on which, coincidentally, we and all other living things depend.

This may sound familiar to you. It's been described in a book called *The Sixth Extinction*. If you are thinking this is a Pulitzer

Prize–winning book written in 2014 by Elizabeth Kolbert,[3] you are correct but only partly so. In 1995, Richard Leakey, famed paleontologist and conservationist, cowrote a book called *The Sixth Extinction: Patterns of Life and the Future of Humankind*.[4] He was one of the first to warn us that human activity was causing a sixth extinction.

Most scientists agree that a lot of factors, not just one, are causing our current mass extinction. We can start with habitat loss, degradation, and fragmentation, as they are probably the worst threats to biodiversity. Just as I explained earlier that our planet is constantly changing, how we use our planet constantly changes too.

Globally, 50% of endemic species of plants and vertebrates are restricted to some 36 biodiversity hot spots covering just 2.5% of Earth's surface. That's pretty cramped quarters. We know that insect populations live the world over, but they're at their highest concentrations in these same hot spots. Recent modeling exercises now predict that the agricultural and economic pressure for land will reduce the natural intact vegetation by another 50% by 2050 in one-third of the world's hot spots.

[3] *The Sixth Extinction: An Unnatural History* by Elizabeth Kolbert, New York: Henry Holt and Company, 2014
[4] *The Sixth Extinction: Patterns of Life and the Future of Humankind* by Richard Leakey and Roger Lewin, New York: Doubleday, 1995

TAKING OVER

We can't seem to stop ourselves from gobbling up land and pushing everything that belongs there aside. We grow our cities and build and build and build. We also mine, tunneling through Earth like moles and disrupting anything and everything that lives in our soil. It's not just the expansion of agriculture but also the intensity with which we use the land. In large parts of Europe, the US, and South America, monocultures cover vast areas of the landscape, creating biological deserts devoid of hedges or ponds where insects could reproduce. These practices eliminate biodiversity and destroy natural habitats.

If we aren't completely destroying these biodiverse areas of land, we fragment them, chopping them up into smaller and smaller pieces. This can be particularly hazardous to insects. Those with low mobility get trapped, isolated, and eventually die out. Species with high mobility may form metapopulations. With so many living things sharing the same small patch of land, there are fewer resources for all.

Air pollution—including chemicals spewing from factories and mining operations—makes things unsafe for insects.[5] Industrial discharge, sewage, agricultural and urban runoff, and increased sediment deposition all foul up freshwater habitats, threatening the insects that live there.

POISON AND POLLUTION

We poison *everything*. We hose our plants down with insecticides, microbial pesticides, and herbicides. The introduction of neonicotinoids in the US and Europe in the mid-1990s has been crippling

---

[5] And to other living things as well, including people

to insects. Only 20% of the insecticide that coats the seeds is taken up by the crop, so the rest remains to be touched or eaten by something else. Neonicotinoids may well be the DDT of the early 21st century with indirect effects throughout food chains. Even fertilizers designed to help our plants grow can be harmful to insects, causing (among other problems) soil acidification.

Wherever we go, we make noise. Have you ever noticed that? We are not a quiet species, human beings. Our cars putt-putt down roads, our planes thunder through the skies, our ships slosh through the oceans. We've long known that excess noise causes adverse effects on human health. Well guess what? A 2019 study from Queen's University Belfast discovered—no surprise—that noise pollution affects the health of practically *all* living things, including insects. This review of more than 100 noise pollution studies turned up some interesting insect data for us.

One 2017 study conducted by the Florida Museum of Natural History found that the noise generated by gas compressors in New Mexico is taking its toll on the local arthropods. Gas compressors, which can be as small as a minivan or as large as a warehouse, extract and move natural gas along a pipeline. These machines operate 24 hours a day, 365 days a year and produce both high and low frequencies. They also emit vibrations into the ground.

Populations of grasshoppers, velvet ants, and wolf spiders as well as cave, camel, and spider crickets all dropped significantly. Because those insects rely on the ability to make sounds or sense vibrations—to find a mate or detect prey—the compressor noise could interfere with information they either receive or send out. Oddly enough, leafhopper populations near compressors surged upward. For them, the noise may act as a predator shield, masking their presence. The bats and birds that usually eat them cannot tolerate areas with excessive noise. The noise is bad for insects but could also trigger a chain reaction along the food chain.

A 2016 study conducted by researchers at Vassar College in New York found that the noise from regular traffic interfered with the mating calls of tree crickets. The males, discouraged that their calls could not be heard over the traffic noise, shortened the length of their songs and inserted more pauses. They conserved energy, but the changes in their behavior means the females are less likely to hear their calls.

We're loud, *and* we're afraid of the dark. Artificial light covers about a quarter of the Earth's surface. Roughly half of the millions of insect species on Earth are nocturnal, so researchers recently reviewed 229 studies that looked at the impact of artificial light at night on insect species.

This review yielded some disturbing insights. A common impact of light pollution is moths flapping around a bulb, mistaking it for the moon. One-third of insects trapped in the orbit of artificial lights die before morning either through exhaustion or being eaten by predators.

Vehicle headlights pose a deadly moving hazard, resulting in an estimated 100 billion insect deaths per summer in Germany alone, where a study was conducted.

More generally, nocturnal insects rely on natural light (the moon and stars) to orient themselves and navigate. Some insects use the polarization of light to find water sources so they can breed because light waves line up after reflecting from a smooth surface. Mayflies, for example, only live for a day and need to lay their eggs in water. If they find a shiny puddle shimmering on an asphalt road, they will lay their eggs there only to have them quickly destroyed by oncoming traffic.

Artificial light can disrupt insect communication. It hinders insects from finding a mate in some species. Fireflies need darkness to exchange bioluminescent signals during courtship.

Light pollution changes the perceived length of the day and night. This affects the development of some juvenile insects, which happens with field crickets. It affects the ability to search for food. Insects that avoid light (like the giant flightless New Zealand cricket, the wētā) will spend less time foraging for food in light-polluted areas.

Insects are important prey for many species, but light pollution can tip the balance in favor of the predators. Spiders, bats, rats, and reptiles have all been found feeding around artificial lights. Such increases in predation risk is likely to cause the rapid extinction of affected species.

What else? It's bleak, but I must carry on. This may sound ridiculous, but we overexploit insects. We don't hunt them down, exactly, but we do overharvest them for a whole host of reasons. We eat them in unsustainable ways. There's a medicinal demand for them to be used in some traditional medicines and treatments.[6] They are illegally trafficked as pets—I'm not kidding about that. There's also a whole illegal export industry for high-demand or ornamental insects that people seek to collect or to use in decorations and jewelry.

We're clumsy. We introduce alien invasive species to places *all the time*. Invasive species are primarily spread by human activities,

---

[6] The commercial value of products based on medicinal insects comprises about $100 million per year.

often unintentionally. People and the goods we use travel around the world very quickly, and we often carry uninvited species with us.

You've probably heard about some of these. The Asian tiger mosquito (*Aedes albopictus*). It was initially found at the Port of Houston in a shipment of used tires from Southeast Asia in 1985. It's a horrible pest that transmits all sorts of pathogens such as yellow fever and Zika virus. How about the emerald ash borer (*Agrilus planipennis*), which may have come to North America in wood packaging material from Eastern Asia in 2002? That thing is a highly destructive beast that kills ash trees by the tens of millions. Or the Argentine ant (*Linepithema humile*). No one even knows how this got to North America. All we know is that it established a megacolony that continues to grow. They are now ranked among the world's 100 worst animal invaders. They displace all native ants and disrupt entire ecosystems.

There are a lot of examples, unfortunately. Non-native species can drive local populations to extinction through predation, competition, or disease. It's not just non-native insects that can cause problems. Non-native plants can too. They can be poor hosts to the local insects, as a food source or as a nesting site. Invasive pathogens can also wipe out native insects. European strains of the fungal pathogen *Nosema bombi* are thought to have resulted in the widespread collapse of North American bumble bees.

OUR CHANGING WORLD

Insects are also struggling with climate change, of course. We hear a lot about climate change. Climate refers to the usual weather of a place. Earth's climate is what you get when you combine all the climates around the world together; it's always changing, but usually the changes take place slowly and can last millions of years. In the last 100 years, our planet has warmed considerably. Our friends at the National Oceanic and Atmospheric Administration (NOAA) estimate that there has been a 2°F (-17°C) increase in average global surface temperature since the preindustrial era (1880–1900). The combined land and ocean temperature has increased at an average rate of 0.07°C (0.13°F) per decade since 1880; however, the average rate of increase since 1981 (0.18°C/0.32°F) is more than twice as great.[7] That maybe doesn't sound like much until you put it into perspective.

Given the sheer size of our planet and the tremendous heat capacity of our oceans, it takes a massive amount of heat energy to raise Earth's yearly average surface temperature by even a little. All that accumulated heat is driving regional and seasonal temperature extremes, reducing snow cover and sea ice, raising ocean levels, intensifying heavy rainfall, altering the timing of when certain plants grow, and changing habitat ranges for plants and animals— expanding some and shrinking others.

We know that insects are very susceptible to temperature changes. They are ectotherms, remember. They can't regulate their own body temperatures and must rely on their environment to warm them up or keep them cool. Which is why people's heads nearly came off in 2018 after reading the published work of a team of biologists studying rain forest insects in Puerto Rico. Bradford Lister, from Rensselaer Polytechnic Institute in New York, and his colleague, Andres Garcia, an ecologist at the National Autonomous

[7] The 10 warmest years on record have all occurred since 1998, and nine of the 10 have occurred since 2005.

University of Mexico, have been conducting research there since the 1970s. The team became worried about the rain forest's insects after noticing that temperatures at two sites had risen by about 2°C (3.6°F) in the intervening decades.

Compared with the 1970s, the surveys from 2011 to 2013 turned up 98% less insect biomass in ground traps, 83% fewer insects in sweep nets, and 65% fewer insects in canopy traps. The scientists also found close to 60% fewer anoles, a family of lizards that eat insects. Lister and Garcia looked at the abundance of other animal populations—canopy arthropods, walking-stick insects, frogs, and birds—elsewhere in the forest. Everything was declining.[8]

Well, holy crap. We're doomed then, aren't we? The planet is getting hotter. The insects are all going to die. We're going to die. Game over. No! Hold up a second. The situation is more nuanced that that. A hotter planet doesn't necessarily mean that all insects are doomed everywhere. No, in fact, some insects may thrive in the heat. Or they may move to find more comfortable places to live, which would change other ecosystems. If we've learned nothing else together so far, we've learned that everything is interconnected. The Lister/Garcia study helps us see that clearly and should drive continued research there and in other places. The only thing we know for certain is that we need to know more. A lot more.

We know damage has been done. But how much, exactly? Remember when I told you about the background rate for extinctions? The normal rate for mammals was one species lost for every

---

[8] Further analysis of their data is aimed at assessing climate damage brought on by Hurricane Hugo in 1989 and the even more devastating Hurricane Georges in 1998—both extreme weather events.

200 years. In the last 400 years, we've lost 89 mammalian species to extinction. Using fairly conservative estimates, a 2015 *Science Advances* study placed the current pace of global extinction to be 100 times the normal background rate. You can understand the panic.

THOSE WE'VE LOST

Which animals have we already lost? When we think of extinct animals, everybody's minds always jump to the dodo bird (*Raphus cucullatus*). Why do we think of that? It's been dead since 1681. None of us remembers it. None of us ever saw it! But there are a lot of things that have become extinct in the not-so-distant past that we either knew or at least have heard of.

There was Martha, of course. Yes, we named her. In honor of First Lady Martha Washington. Only, this Martha wasn't first; she was last. The very last. Martha was the last passenger pigeon (*Ectopistes migratorius*) on Earth. She died in captivity at the Cincinnati Zoo in 1914.[9] Passenger pigeons were endemic to North America and ranked as the most abundant bird—so numerous they tallied in the billions. They must have been good to eat because we hunted them to extinction.

There was Lonesome George. He was the last Pinta Island tortoise (*Chelonoidis abingdonii*) and died in the Galápagos Islands in 2012 after reaching the centenarian mark. Pinta tortoises were over-exploited by whalers, fur sealers, and others in the 1800s.

Other species died out without a named poster child we could relate to. In 1924, the last of the California grizzly bear (*Ursus arctos californicus*) was spotted. In 1936, the last thylacine (*Thylacinus*

---

[9] She may have been 29 at the time of her death, which isn't bad for a pigeon.

cynocephalus) died in captivity. They were a carnivorous marsupial native to the island state of Tasmania, New Guinea, and the
Australian mainland. They evolved about two million years ago,
but hunting, habitat loss, disease, and competition from domestic
dogs all brought about their downfall. In 1985, we saw the end
of the northern gastric-brooding frog (*Rheobatrachus vitellinus*)[10]—
from habitat destruction and disease. We lost the golden toad
(*Incilius periglenes*) of Costa Rica in 1989 to climate change. The
eastern cougar (*Puma concolor couguar*) and western black rhinoc-
eros (*Diceros bicornis longipes*) both left our mortal coil in 2011.
The Japanese river otter (*Lutra lutra whiteleyi*) followed in 2012. So
many losses.

If you're looking for numbers, try the International Union for
Conservation of Nature's Red List of Threatened Species, established in 1964. It has evolved to become the world's most comprehensive information source on the global conservation status
of animal, fungi, and plant species. They will tell you that, of the
species that have been assessed (only 28%), more than 38,500 species are currently threatened with extinction. While that's the best
guess, it's probably nowhere near accurate. How many of those at
risk are insects?

You've probably heard less about the insects we've already lost,
I'm guessing. Like the Caribbean monk seal nasal mite (*Halarachne
americana*). Following the 1952 extinction of the Caribbean monk
seal—the nose in which the mite lived—the mite was unable to

---

[10] A cooler frog there never was! Gastric-brooding frog tadpoles developed in
their mother's stomach for six to seven weeks. When they were ready to be born,
their mother burped them up. The young would shoot out of her mouth and hop
away.

adapt and subsequently went extinct. *Oh well. A nasal mite*, you might say. *The world can probably get by without THAT, right?*

What about the levuana moth (*Levuana iridescens*)? Its only crime was that it loved coconuts. Unfortunately for this moth, coconuts are a major cash crop on the island of Fiji. The poor thing was the target of an intense eradication campaign in the early 20th century, which sadly succeeded. Because Fiji is such a small island, the moth had nowhere to go as it was being hunted down.

The Rocky Mountain locust (*Melanoplus spretus*), which we talked about in Chapter Six, was to the insect world what Martha and the other passenger pigeons were to the bird world. During the late 19th century, passenger pigeons numbered in the billions while the Rocky Mountain locusts numbered in the trillions. The locust succumbed to agricultural development. The last credible sighting was in 1902.[11]

The Madeiran large white (*Pieris brassicae wollastoni*) was a stunning black-and-white butterfly as big as your hand. It was last collected off the coast of Portugal in the late 1970s. Gone now as a result of habitat loss. As is Sloane's urania (*Urania sloanus*)—a Jamaican moth of unusual colors . . . iridescent red, blue, and green. It flew during the day rather than at night, which is a common habit of tropical moths. It was doomed by the conversion of Jamaica's rain forests to farmland.

Then there's the Xerces blue (*Glaucopsyche xerces*), which had to go extinct in full view of butterfly lovers across the United States. It was considered to be the first American butterfly species to become extinct as a result of loss of habitat caused by urban

---

[11] I don't feel bad about that one. You know I hate locusts. I know this makes me a bad, bad person. At least I'm honest.

development. The last of its kind was spotted in the early 1940s on land that is part of Golden Gate National Recreation Area.

These are a handful of insects I can rattle off, but that's hardly a comprehensive list. One of the earliest high-profile, data-driven studies that got people's attention came in 2004 by J. A. Thomas et al. in *Science*, which documented the rate of butterfly decline in Great Britain. Soon after, the world was greeted with studies on colony collapse disorder and the honey bee's struggles with varroa mites. In 2007, the National Research Council of the National Academy of Sciences released a high-profile report on the status of pollinators in North America, which brought attention to bees, monarch butterflies, and other pollinators that appeared to be in decline.

Rodolfo Dirzo et al.'s 2014 "Defaunation in the Anthropocene" was the first meta-analysis to report global insect losses for beetles, dragonflies, grasshoppers, and butterflies. Across 16 studies, insect populations had declined by 45% in the last four decades.

Then came a shocking study from the Krefeld Entomological Society near Düsseldorf, Germany. In 2017, a team of researchers reported a decline of *more than* 75% in insect biomass across 63 nature areas in Germany between 1989 and 2016. The reasons for that decline aren't clear, but everyone was unnerved by the data. The rate of loss was much higher than anyone expected. Shockingly, the areas that were studied were protected areas—most of them were managed nature preserves. Researchers took notice of this study—it quickly became the sixth most discussed scientific paper in 2017, according to Altmetric, a website that tracks such statistics.

The news seemed bad everywhere you looked, in large studies and small. Researchers reported dramatic declines in moths attracted to light traps in England. Another study showed a 14% decline in ladybugs in the US and Canada from 1987–2006. Ecologist Toke Thomas Høye of Aarhus University in Denmark studies muscid flies in remote Greenland and found an 80% drop in their numbers since 1996. In *Greenland*. These flies are 300 miles from civilization yet they're in decline. A 2010 international gathering of firefly experts reported downward trends.

Study after study shows insect declines and the ramifications within those ecosystems. People who studied fish found in one study that fish had fewer mayflies to eat. Another recent study suggests farmland birds that depend on a diet of insects in Europe have declined by more than 50% in just three decades. Eight in 10 partridges are gone from French farmlands. Nightingales and turtledoves are down by almost 80%. They have nothing to eat because the insects are gone.

A 2018 US census found that the population of monarch butterflies fell by 90% in the last 20 years. *That's a loss of 900 million individuals.* Let that sink in for a minute. The rusty-patched bumble bee dropped by 87% over the same period. Anecdotal evidence from many parts of the world also indicates insect declines. The 2019 work of Francisco Sánchez-Bayo and Kris A. G. Wyckhuys was met with an uproar from the scientific community after it concluded that 40% of insect species might be at risk for extinction in coming decades.

IS IT THE APOCALYPSE?

Headlines in mainstream media were screaming "Insect Apocalypse" and "Insect Armageddon." Most of the studies paint a far more complex picture than these jarring headlines. The trouble with insects, in some ways, is that we have a very poor understanding of our insect baseline. We simply don't know what we might be losing because we haven't systematically looked. That's why, in recent years, we've been seeing megastudies that cobble together as many smaller studies as can be found in the hope of creating a fuller picture. What every insect needs is a Dr. Arthur Shapiro. But only the butterflies have Dr. Shapiro.

Dr. Shapiro is an entomologist at the University of California, Davis. In 1972, he began walking transects of the Central Valley and the Sierras, counting butterflies. Today, from late spring until the end of autumn, he visits each of 10 locations every two weeks. He's in the field about 260 days a year and covers about 24 kilometers (15 miles) a day. His mission is to record and identify every butterfly he sees. His project was planned for five years. It has been running for 49 and continues to this day. Dr. Shapiro has single-handedly created the longest-running butterfly monitoring project in North America.

Dr. Shapiro cuts an intimidating figure in some ways. His face is lined and weathered from all his time in the field. He looks, one might think, a bit feral. His hair is not unlike the mane of a lion. Full and curly, it juts out from his head only to meet up with a long and flowing beard. You think of a lion as a fierce apex predator, but when Dr. Shapiro speaks, his voice is warm, friendly, and oddly supportive. As if all his years of being a professor are so deeply ingrained that his instinct toward you is merely to help you understand whatever it is you're grasping at. I've been trying to grasp a lot lately—about research, about conservation—and I'm hoping he can help. What's happening with the butterflies? What's happening with all insects? What data sets do we need, and who needs to be doing the research? How should such programs be structured to ensure success? How do we even measure success?

"Well, I'm not sure what to tell you," he says kindly. "It's really tricky." Dr. Shapiro doesn't have all the answers for me, but he can tell me a lot about his own work with butterflies. His data illustrates the overall decline in 163 species of butterflies from sea level to 2,500 meters (8,200 feet) of elevation. He's seen species that were extremely common all but disappear.

His study was never really intended to reveal this kind of information. It was supposed to be a limited study that looked at the relative factors that determine the timing of butterflies' life cycles.

"The British Butterfly Monitoring Scheme[12] was designed from the beginning to go on forever. But mine was not," he says. "It was only because California has had such extremely erratic weather and the data were so exciting that it's become open-ended. It was just unthinkable to stop gathering data."

As Dr. Shapiro describes his work to you, you feel a rising sense of righteous indignation. Can no one help this man? Does he really have to do it all by himself? He doesn't see it that way at all, though. First of all, he's had help. He's trained countless students over the years. Second, because he's done this work for so long—in the same way with absolute consistency—he knows his data are sound and rigorous and therefore valuable. Third, he says he's cheap.

"What I do is exceedingly cheap," he tells me. "If I had to rely on grant funding on the usual short-term cycle, I probably would never have gotten into this in the first place. It's just me doing it,

_____
[12] Officially the United Kingdom Butterfly Monitoring Scheme, ukbms.org

and I'm not flying to Venezuela for the weekend to get data. So it's cheap."

Well maybe. But still. It seems like an incredible burden, this work.

"I'm not going to continue forever, but at least we've gotten out so much data and so much analysis that what we have done could be a baseline for others and maybe an indicator of what to do or what not to do."

Not everyone can devote themselves to their work as Dr. Shapiro has. Are there things that regular people can do to help? If we can't walk the transect with him every day, can we do something else? Plant milkweed or something?

"Let's see . . . How do I put this . . . ? There's no shortage of milkweed. There's lots of milkweed; there's no butterflies to breed on it. Therefore, planting milkweed will not do any harm, but it won't do any good."

"What *can* we do?"

Dr. Shapiro gives me a good list: You can plant good nectar sources for the butterflies because those are often in short supply. You can minimize your use of pesticides. You can encourage your local highway department to minimize or, if possible, eliminate the use of herbicides for vegetation management. And you can try to reduce your carbon footprint.

Those all seem like reasonable things, but are they enough? I hate to use the E word, but it seems fairly clear that we are in the middle of an extinction event, and this time, our insects are at risk. A total extinction of insects means a total extinction of life on Earth. That's hard to conceive. And, honestly, we're not there yet. Yet. But what if we're facing extinction with a lowercase *e*? Scientists are talking more and more about functional extinction. Functionally extinct insects still exist on our Earth, but they're no longer prevalent enough to affect how an ecosystem works. Their absence causes hiccups. Animals that eat the insects, for example, have to eat something else. Or, they leave the ecosystem to find their prey elsewhere. The more these connections are lost in an ecosystem, the more unstable it becomes.

LOCALIZED EXTINCTION

Sometimes, an extinction is localized. You remember the extinction of the St. Helena giant earwig (*Labidura herculeana*) in 2014, right? (*The what what?* you say.) The giant earwig was endemic to St. Helena Island in the Atlantic Ocean. During the day, our friend the earwig burrowed under rocks. At night—but only after a rain—it would crawl out looking for food. Then along came people. They removed the earwig's rocky shelters because they needed building supplies. With the people came earwig-eating rats, spiders, and centipedes. Slowly the earwig began to disappear. Oh, there were searches—entire (two-man) expeditions searched and searched. Several times. But nothing. Finally, the giant earwig was officially declared extinct.

You missed that one? Maybe it doesn't feel important to you because it doesn't seem to affect your life in any meaningful way. I get that. The loss of the giant earwig might not even affect the ecology of St. Helena all *that* much, right? The rats got a little hungrier maybe. Except that the St. Helena giant earwig is gone from the island and from Earth's face forever. That's a localized extinction and a loss of diversity. It may not look like it matters, but it does. Here's how: A 2013 paper in *Nature*, which modeled both natural and computer-generated food webs, found that a loss of even 30% of a species' abundance can be so destabilizing that other species start to go extinct. It's these common species, because of their abundance, that power the living systems on our planet.

I'm going to tell you a story in the next chapter. It's about an insect you've probably never heard of that became extinct but not really. This story is about the Herculean efforts that a dedicated few made to ensure this insect not only survived but also thrived. I'm hoping you'll see how and why it all matters.

8.

HOW IT MATTERS

IT WAS IMPOSSIBLE not to look at his thighs. Not Dr. Ben Price's thighs. As far as I know, Dr. Price's thighs are perfectly proportional for a man of his height and weight. No. I mean the other pair of male thighs in the room. The pair that belonged to the Lord Howe Island stick insect.

I wondered if my mom was noticing them too. It was hard not to. She and I had come a very long way to see those thighs and the insect attached to them. Well, that's not exactly right. We came to London to celebrate my birthday, but during the planning of the trip, I told my mom that I needed to make a quick stop at the Natural History Museum in London to meet Dr. Price, the senior curator in charge of small orders in the Insect Division.

It's true that my mom does not love insects. I believe the word she usually uses is "icky" right before she asks if it's okay to have my dad "kill this one." (No, Mom. Not okay.) That said, my mom has an insatiable curiosity, so when I offered to let her sit this one out, she looked at me like I had grown a second head. The Natural History Museum has one of the oldest and most important entomology collections in the world. It's 34-million insects and arachnids strong. It's not hyperbole to say that there were literally *millions* of insects that Dr. Price could show us, but of all the insects he *could* show us, he was rather keen to show us this one. It is, after all, the rarest insect in the world.

You've not heard of the Lord Howe Island stick insect before? That's okay. It's quite spectacular, though. *Dryococelus australis*, if you want to be formal about it. Although it's got quite a few clever nicknames such as "tree lobster" and "the walking sausage."

It looks a bit like a praying mantis disguised as a cockroach. It's about as long as your hand with a shiny, reddish-blackish exoskeleton. The males are smaller than the females, except in the thigh department. The males' thighs—the size of which are believed to attract females—are meaty with large spines jutting off them. They are a sight to behold. I'll pause for a moment while you google it.

Got it? Good. I'm not wrong, am I?

The specimen my mom and I were looking at was collected from the island in 1855,[1] and its thighs were so chunky, the lady stick insects must have noticed. I don't mean to body shame the thing, but there's no delicate way to put it: These are Thunder Thighs.

If you can look past the thighs—a very difficult thing to do— and listen carefully, you'll learn the amazing story behind this insect. The Lord Howe is something of a Lazarus, you see. It was declared extinct and only came back from the dead when it was accidentally rediscovered. It was living in a place it shouldn't have been, and it was brought back by a team of scientists who risked their lives to do so.

---

[1] It was first described by a French missionary, explorer, and scientist named Jean Xavier Hyacinthe Montrouzier.

ORIGIN STORY

The Lord Howe Island stick insect came from—of course you can guess this—Lord Howe Island. That's a tiny Australian island in the Tasman Sea known for its sandy beaches, subtropical forests, and apparently stick insects. These types of insects are also called phasmids from the order Phasmida.

The Lord Howe phasmid was living on its island paradise until disaster struck. It was a dark and stormy night on June 15, 1918, when the steamship *Makambo* "came to grief" off the coast of Lord Howe Island. No, actually, I made that up. It wasn't dark and stormy at all—it was a clear, moonlit night with calm seas. The *Makambo* did come to grief, but it was the captain's fault, not the sailing conditions'. Not that I blame the captain entirely. He was having a pretty rough time of it. He had whooping cough, which, as anyone in 1918 could tell you, was no picnic. It's not as common today as it was then because we vaccinate against it. But it is wildly contagious and characterized by a severe, hacking cough followed by a high-pitched intake of breath that sounds like "whoop!" The cough can lead to vomiting or, in severe cases, unconsciousness.

It was just such a condition that dear Captain Wetherall of the *Makambo* found himself in that night. He was belowdecks when he should have been on deck barking orders to his crew to avoid hitting the reef off Lord Howe Island. By the time his first officer found him inert on the floor, it was too late. The ship struck the reef with such force, the crew was collectively knocked off their feet.

It became quickly clear that the *Makambo* was taking on water and that they had a situation. In the ensuing panic—and there was a lot of that, I can assure you—the crew tried launching the lifeboats to evacuate the passengers. They also tried jettisoning some of the cargo, namely some large crates of fruit and copra.[2] None of that went according to plan, though. The ship began listing to one side, then, worse, rolling in the swells. The ropes got hung up on one lifeboat, leaving it dangling precipitously off the side of the ship.

By now, the islanders, having heard the ship's siren and having seen the foundering ship, came to retrieve the passengers. No one paid much attention to the jettisoned cargo.

Those crates ended up on shore, though, and with them came tiny stowaways: *Rattus rattus*—a name so terrible, we say it twice. They are also known as ship rats for their tendency to thrive on seafaring vessels. There's much to be said about *Rattus rattus*, but for our purposes here, we can just declare them evil and be done with it because they are certainly no FOSI (friend of stick insects).

You're no doubt wondering how all this—the *Makambo*, Captain Wetherall and his whooping cough, and a passel of manky rats—could possibly relate to the Lord Howe Island stick insect. I'm getting to that . . .

As an island, Lord Howe didn't have much to speak of in the way of predatory mammals. When the sea-soaked ship rats came slithering ashore, there was nothing on the island to prevent them

---

[2] You might know what copra is, but I had to look it up. Apparently, it's the dried meat, or kernel, of the coconut. Coconut oil is extracted from it, making it an important agricultural commodity. Who knew?

from flourishing. These little stinkers are known as generalist omnivores. In laymen's terms, that means they aren't too picky about what they eat, be it seeds, fruit, stems, leaves, fungi, or a variety of vertebrates and invertebrates. And so the rats ate everything in sight. Sadly, they developed a special fondness for the Lord Howe Island stick insect.

The *Makambo* lay aground for another nine days. Those rats that hadn't already jumped ship aboard a crate could all but walk ashore while the ship was beached and being repaired. Nine days is more than enough time for a rat colony to evacuate a sinking ship and set up shop elsewhere.

The rats were voracious. They were also prolific. Between stuffing themselves sick with stick insects and other things[3] and breeding like maniacs, the island didn't stand a chance. It should be noted that at one time, the island was crawling with stick insects. They were so plentiful, fishers used their chunky back legs as bait on their fishing lures. But by 1920, not a single Lord Howe Island stick insect could be found.[4]

A DISCOVERY AT BALL'S PYRAMID

The story doesn't end here, of course. Not by a long shot. Those rats would cause no end of trouble for the island, but the stick insect was not, in fact, extinct.

Fast forward 44 years. Travel 23 kilometers (14.2 miles) southeast of Lord Howe Island to a place called Ball's Pyramid. This

---

[3] Five species of land birds—thrush, warbler, starling, fantail, and silvereye—all disappeared during that same period. The gluttonous rats ate them into extinction. Seabirds that once bred on the main island were forced to nest on the outlying islets. Many other species (including the island's lizards, snails, and beetles) became threatened as a result of the rats.
[4] They were eventually classified as extinct in 1986.

rocky pinnacle is what remains of a seven-million-year-old shield volcano. It pierces the ocean like a shard of broken glass. At 562 meters (1,843 feet) tall, it is the largest volcanic sea stack in the world. This island is guarded by a reef of coral rock and rough, turbulent seas. It is an inhospitable place. The rock is at the mercy of blistering heat, salty waves, strong winds, and frequent storms. Beyond ground cover, only sparse patches of a spindly bush called a *Melaleuca howeana* populate the rock. The Pyramid is difficult to reach and challenging to climb.

In 1964, a climbing team from Sydney attempted to summit. It was an unsuccessful bid, but one climber from the group, David Roots, saw something interesting on the way up. He took a photograph of a weird-looking, albeit dead, stick insect. It would be a few years before anyone took a hard look at that photo, but when they did, they saw what looked to be the Lord Howe Island stick insect.

Ball's Pyramid was summited in 1965. On a 1968 climb, Jim Smith (a climber and zoology student) found two partial exoskeletons that also appeared to be Lord Howe Island stick insects. Now things were getting interesting. The insect world was all abuzz. If there was any possibility that the phasmid might somehow be still alive, then Ball's Pyramid was a place that needed protecting.

It was thought that future climbers might pose a risk by tracking some nasty bit of microscopic whatsit that could wreck the delicate constitution of the phasmid should one happen to be wandering about up there. Oh, and there was something else about the climb being too dangerous . . . something something . . . risk of sudden death and such. A climbing ban was put into place, and a hefty fine would be assigned to anyone brazen enough to flout it. Seemed reasonable enough. Public safety and all.

Except that the ban did nothing to dissuade thrill-seekers and earnest climbers alike from petitioning the right to climb it, mostly under the guise of a scientific expedition to search for the phasmid. *Not sure those stick insects are there? Let me go in search of them. If I see any on my way to summiting, I'll be sure to let you know.* Uh-huh.

When requests of this nature were made, they were received by the New South Wales Office of Environment and Heritage, and they often made it to the desk of Principal Research Scientist David Priddel. Priddel was easily skeptical of the requests. How serious could they be about the phasmid? The proposed expeditions rarely included an entomologist or even a natural scientist. He figured most requests were less-than-clever ruses by climbers determined to get to the top of Ball's Pyramid.

Priddel held no illusions that the phasmid was there. It didn't make sense. The barren rock of Ball's Pyramid offered little in the way of vegetation. It lacked the humidity the stick insects had enjoyed on Lord Howe Island. How would they have gotten there anyhow? Lord Howe and Ball's Pyramid have never been physically connected. Without a land bridge, how could the wingless Lord Howe Island stick insect *get* there?

No. The whole thing seemed too far-fetched. Convinced the phasmids were truly extinct, Priddel decided to prove it once and for all. He mounted his own expedition. He just needed the right team.

A NEW EXPEDITION

Dr. Nicholas Carlile seemed an obvious choice. Dr. Carlile worked with Priddel. He specialized in island and sea bird ecology. He was young, enthusiastic, and dedicated. Margaret Humphrey, an entomologist associated with the Australian Museum in Sydney, came highly recommended. They would need her expertise. The fourth member of the crew was Stephen Fellenberg. Fellenberg owned a private insect-breeding company called Insektus. He had valuable experience with phasmids.

That set the core of the team of scientists, but one more team member was needed: Dean Hiscox. Hiscox was one of Lord Howe Island's most experienced rangers. He was an expert climber and had practical know-how in cliff rescue, should it come to that.

For Fellenberg, it did. But not on Ball's Pyramid. While the team waited for the all-clear to go to Ball's Pyramid, Fellenberg went to search Mount Gower on Lord Howe Island. Gower is the island's highest point and is covered in thick grasses. It was one place the rats might not have reached. Maybe the phasmid lay hidden there. Alas, he found nothing but heartache after injuring both knees on the climb. His hopes to join the Pyramid expedition were dashed. The team was down to four now.

The trip from Lord Howe Island to Ball's Pyramid is a stomach-churning journey through the choppy waters of open sea, after which, you're in for a real treat on the landing. There's no beach to speak of . . . nowhere to land. But it's not like you can get your boat closish then dive in and swim the rest of the way to shore. The water is teeming with sharks.

To land safely, you need good timing, good seas, and an experienced skipper. The team found that in local boat owner Clive Wilson. Wilson brought the boat as close to the rock ledge as he dared. The tiny vessel rose and fell on 2-meter swells. The trick was to wait until a strong enough wave crested to bring the boat level with the ledge so the team members could leap across. There wasn't time to keep one foot on the boat while stepping onto the rock. It had to be done in one solid leap. If the timing was off by even a second, one of two things would happen: You'd fall into the drink and become a shark tater tot, or you'd fall into the drink and be smashed against the rocks. Neither choice was a winning outcome.

It took more than an hour for the team to safely off-load themselves and their gear. They had planned to stay overnight but packed enough provisions for three days. Conditions on Ball's Pyramid had a way of rapidly changing. If the weather shifted and they got stranded, they'd at least be well-supplied.

It was late morning by the time they secured their campsite, stowed the extra gear, and began their trek. They had identified a spot about a third of the way up that they wanted to survey. It was called Gannet Green, and it was where one of the island's only pockets of vegetation lay.

Before the team could begin their ascent, they first had to inch along the base of the Pyramid. There was a 300-meter section that skirted above the water line by a scant 2 meters. The footing here was wet in some sections, dry and crumbly in others, and barely a few centimeters wide. While Priddel and Humphrey clung to the rocks by their fingernails, Hiscox and Dr. Carlile scampered ahead to hastily install fixed ropes over the more difficult passages. In retrospect, these ropes were probably more for their mental comfort than to serve as actual supports should someone wobble and fall.

If this inching along wasn't harrowing enough, the team also had to contend with seabirds by the thousands. Clouds of birds—masked boobies, white-bellied storm petrels, red-tailed tropic birds—took to the skies around the Pyramid. A birder's paradise to be sure. When the team wasn't skirting around occupied masked booby nests on the ground, they were ducking sooty terns who were dive-bombing the intruders for encroaching on their turf.

The sound of birds was deafening but a minor inconvenience compared to the poop. Bird poop rained down liberally on the travelers during most of their trek.

The task at hand, of course, was to search every possible area along their ascent for any signs of the phasmid. Remember, the team was out to prove a negative: that the phasmid was *not* there. They never expected to see one, but they had to rule out seeing any *evidence* of one as well. That meant searching for eggs, dead bodies, body parts, or exoskeletons that had been shed.

The day was hot, windless, and fairly miserable. By the time they made it up to Gannet Green, they were exhausted and had discovered nothing.

Still they carried on with the next phase of the search. They systematically beat bushes with sticks and held trays underneath on the off chance something might be dislodged. They also sifted soil, searching for eggs.

It was pretty late in the day when they came to a halt. Aside from a few crickets and grasshoppers, they still had found nothing. Dehydrated and sunbaked, Priddel called off the search. They regrouped and began the long descent to their camp. Dr. Carlile was in the lead when they were about 70 meters (about 230 feet) above sea level and passing a small clump of vegetation. It was a *Melaleuca howeana* shrub, which is a type of tea tree. They had passed this same shrub on the way up and took no notice of it.

But coming down, Dr. Carlile trained his eyes on it and saw

a large pile of frass. Frass, as you know, is insect poop. When I say it was a large pile, you have to understand that insect poop—even of the phasmid variety—is rather small. About half the size of a Tic Tac. To spot even a large pile of Tic Tac halves is to have remarkably good eyesight. But spot them, he did, and he quickly called Humphrey over to have a look. If anyone would have a professional assessment of this pile, it would be the entomologist. It was an impressive heap, she conceded, but difficult to say with any certainty who the pooper might have been. It could have been a phasmid, but it could just as equally have been, say, a robust cricket.

The team continued their descent, each lost in their own thoughts. Back at camp, Dr. Carlile was dodging another bird poop storm while trying to prepare the evening meal. Priddel sidled up to him for a quick chat. The goal of the entire expedition was to prove that the Lord Howe Island stick insect did not exist in this place. Yet the poop heap now left room for doubt. *The adults*, he explained, *are nocturnal.* Yes, yes, Dr. Carlile knew that. *I think we have to go back up there to search. At night.*

Priddel carefully explained that he couldn't ask Dr. Carlile to go up at night because he himself couldn't manage a night trek. Dr. Carlile wasn't really listening to that. He stopped listening after the word *nocturnal.* He quickly volunteered to go and volunteered Hiscox to accompany him ("I'm sure Dean will be keen to do it" were his exact words.) Luckily, Dean *was* keen to do it. Around 9:30 that evening, fitted with headlamps, the two men began to retrace their steps up Ball's Pyramid in the dark.[5]

---

[5] Dr. Carlile's philosophy about night climbing: "If you can't see the drop, it doesn't bother you." Ha ha. Riiiiiight . . .

A LONG, DARK TREK

It was slow going, but the men finally reached the Melaleuca bush.

"We saw a big cricket just before we got to the bush," Dr. Carlile said, "and we thought, 'Oh bugger it. It's a no-go. It's just cricket frass.' And then we got to the bush and got the shock of our lives. We became the first people in 80 years to see a live [one]."

In the light of their headlamps, they could quite clearly see two, shiny black phasmids staring back at them. The men could hardly contain their glee. The bush that the phasmids had chosen was fairly close to the cliff face. Careful to avoid falling into oblivion, the men did a cursory inspection and discovered a third, younger phasmid. It was too incredible to believe.

They stood watching the insects for a few minutes, trying to decide their next move. They were not licensed to bring down any live specimens from the Pyramid. But did that rule apply to an insect that had already been classified as extinct? Surely no one would miss a thing that was already gone . . . But they had no idea how many phasmids were actually there and they didn't want to take the only three left in the world.

At the same time, they needed to document their discovery. This was pre-digital, pre-smartphone days. Dr. Carlile had an Instamatic film camera that had birthday shots on it that he hadn't gotten developed yet. There were three shots left. *Click, click, click.*

Taking one last look at their remarkable find, they made their way back down to camp. Priddel and Humphrey were sound asleep when Dr. Carlile crept up quietly and whispered: "We found it. We found it."

There was no sleeping after that. The tiny group huddled together, brewed some tea, and talked until the early hours about their find. Humphrey had had her suspicions. She'd been impressed by the poo but didn't want to get anyone else's hopes up.

Priddel was agog. How could it be possible? Ball's Pyramid was isolated. There was no land bridge. How could a flightless insect get here? Did it hitch a ride with fishers? Swimming was out of the question. What about being carried by birds? There's an interesting idea. Not the insects themselves, maybe, but the *eggs*.

Follow me here for a second . . . The egg of a Lord Howe Island stick insect is about the size of a pencil eraser. It's shaped like a mini barrel. On one end, it has a lumpy bit called a capitula. Neat word, huh? Anyway, these eggs look an awful lot like seeds. Some seeds have the same sort of lumpy bit. In the plant world, it's called an elaiosome. It's filled with fat, which attracts ants. The ants drag the seed back to their secret, underground ant-lairs for a snack. Tucked safely in the ground, the seeds eventually germinate. Ants, it so happens, are fooled by phasmid eggs. They treat them the same way they would a seed. So the phasmid eggs end up hatching under the soil, protected.

Seeds are also eaten by birds. You'll see where I'm going with this. In addition to the little doodly-doo on the end, phasmid eggs are also coated with a tough material called calcium oxalate.[6] It doesn't dissolve easily. A team of Japanese researchers wondered if a phasmid egg could survive if a bird ate it. So they fed phasmid eggs to brown-eared bulbuls. Guess what? Some of the eggs survived. I'm not saying that's what happened on Ball's Pyramid. But I'm not *not* saying it either.

Back on the Pyramid, the team finally caught a few hours of sleep. The next morning, they rose and climbed back up to their new favorite bush. It was unclear where the phasmids might be

---

[6] It's the same stuff humans find in kidney stones. Interesting, no?

hiding during the day, but Humphrey uncovered two phasmid eggs from the leaf litter beneath the Melaleuca bush. They discovered the bush was healthy because its roots were being fed by water seeping out of the rock face. The plant's leaves revealed signs of grazing.

The weather on Ball's Pyramid was shifting, though, and it became clear the team needed to go back to Lord Howe Island quickly. Priddel's thoughts were occupied by the great responsibility that now rested on their shoulders. They could not say for certain how many phasmids lived on the Pyramid, but they could definitively say that they did live there. It was also clear that the environment, as well as the very existence of this species, was exceedingly fragile. He feared that news of their discovery might attract the wrong sort of attention. Unscrupulous insect collectors might stop at nothing to add the Lord Howe Island stick insect to their collections.

A NEW THREAT

Before returning to the island, Priddel swore the others to secrecy as to the exact location of the bush. They all agreed. Having that settled, his next thoughts revolved around establishing a captive breeding program to bring this non-extinct extinct species back to life.

The task would prove daunting. What did they really know about this phasmid? Not much. A close review of the scholarly literature[7] told them it was nocturnal. Sort of. The adults were, anyway. On Lord Howe, they hid during the daytime in humid

---

[7] A single paper on the Lord Howe Island stick insect was written by Arthur Mills Lea in 1916. Lea was an entomologist whose main interest was in beetles. But he wrote a paper on a longicorn beetle that included his observations on the stick insect as well.

places like the hollows of banyan trees. No one really knew what they ate. Melaleuca leaves, ostensibly, but even that was a guess. That's what they were chowing down on at Ball's Pyramid, but to be fair, outside of some scrubby ground cover, that's kind of all that Ball's Pyramid has to offer. No one knew if that was their preferred food source or if they were just making do. They found eggs buried in the leaf litter, but they knew next to nothing about the gestation period of the eggs or what type of soil was best for them to gestate in.

Was it even safe to try developing a captive breeding program? They didn't have a proper head count from Ball's Pyramid. They had only spotted three live phasmids, all of which were female,[8] and the two eggs. Were there more? Were there enough to safely take some off the Pyramid?

In 2002, Hiscox and ranger Christo Haselden were sent back to Ball's Pyramid on a recon mission. Their sole purpose was to do an official head count at night. During their two-hour survey, they found 24 phasmids on Gannet Green—both males and females. They also found more eggs. It would seem that the world's total population of Lord Howe Island stick insects was confined to six small Melaleuca shrubs on a rocky ledge about 30 meters by 10 meters. The population was small but viable. Barely.

The rangers discovered a new threat, though. Because the stakes weren't already high enough. Nooooooo. Let's add one more thing to complicate matters. The danger took the form of an invasive

---

[8] You might be wondering about this. This species is capable of reproducing asexually by a process called parthenogenesis. Unfertilized eggs hatch into females, so reproduction can happen without the presence of males, which is a neat little biological trick that stick insects can fall back on when their species is low in numbers.

plant called a morning glory. I'm not kidding. And yes, it is scarier than it sounds.

This vine was entwining itself around some of the Melaleuca bushes, choking them. Without the Melaleucas, the phasmids would starve.

While all this was sinking in, news of the team's discovery had the scientific world—at least among insect lovers—all ablaze. Many learned men and women refused to believe that the Lord Howe Island stick insect was alive and kicking. Not even Dr. Carlile's photos could persuade them. *Sure, it LOOKS like the Lord Howe*, they said. *But the thighs are smaller than the Lord Howe specimens we have in museum collections. So, really, this thing on Ball's Pyramid must be a different species.*

What could it possibly matter? you might wonder. It does. Here's why. The ultimate goal of animal conservation is often not just to bring the animal back from the brink of extinction. It's also to release captive-born creatures back into their native habitat. To do so, you have to make sure you're working with the original species and not a hybrid or different species. You know the original species can survive in that environment because that's where it came from. If you introduce something new into an environment, you could face negative or, at the very least, unexpected consequences.

The technology to conduct DNA testing on specimens that had been collecting dust since the 1850s wasn't all that strong back in 2001. Until the technology got better (it did), Priddel was determined to save whatever it was that WAS living on Ball's Pyramid.

To build a captive breeding program, he collected two breeding pairs. The amount of paperwork and red tape to be waded through to obtain permission to collect the phasmids could fill a book. Priddel needed the patience of Job to get through it all.

Almost immediately, he began working with Fellenberg, the phasmid expert from Sydney, and Patrick Honan, the head keeper of invertebrates at the Melbourne Zoo. It was Honan who created a carrying system to safely transport the phasmids from Ball's Pyramid. It was a travel tube fashioned from a PVC pipe.

Led by Hiscox, a team of rangers made the ascent and collected the four phasmids. You can almost picture the three of them, navigating the perilous rock in the dark with their headlamps, trying to bag good specimens. It took more than two hours of searching, but they accomplished their task. One breeding pair was given to Fellenberg in Sydney. The other pair was given to Honan in Melbourne. The future of the species now rested in their hands.

THE PATH AHEAD

Since almost nothing was known about these creatures, Honan made a lot of guesses. He constructed a glasshouse for Adam and Eve—yes, he named them. It was adorned with potted Melaleuca plants. He cranked up the humidity to try to approximate Lord Howe Island conditions. He fashioned a number of hidey-holes for Adam and Eve to take up residence—a ball made out of coconut leaves and a log with holes drilled in it. In the end, the pair chose a fixer-upper made of an old finch box. It was about the size of a shoe box with a hinged lid and a hole in the side. During the day, Adam laid next to Eve, draping three of his legs over her protectively.

To create a body of knowledge, Honan needed to record all direct observations. Unfortunately, anything meaningful took place at night. The now-nocturnal Honan kept detailed notes, jotting down his observations in five-minute intervals. Within the first week, he was able to categorize the phasmids' behaviors into four primary actions: Eat. Mate. Move around. Sit still. That last nonaction action seemed especially popular.

Before long, Honan observed Eve tapping her abdomen on the floor of the greenhouse. With all the insect mating going on in between the eating and the sitting, he thought she might need to bury her eggs. He quickly produced a tray of sand, and he was not disappointed. Eve deposited nine eggs in the sand.

Meanwhile, Fellenberg was having a bit of luck in Sydney. The female in his pair laid 21 eggs in the first two weeks. These were collected and buried in various and sundry substrates—sand, vermiculite, peat, and a mix of sand and peat. No one knew exactly what might be best, so they tried a bit of everything.

Then catastrophe struck. Twenty-nine days after being taken into captivity, Fellenberg's male phasmid up and died. The female followed suit the next day. Fellenberg was devastated.

Now all the pressure rested on Honan. Yet things with Honan had taken a turn as well. Eve stopped eating. Something was wrong. Adam seemed unphased, but Eve was clearly in crisis. What could be causing her illness? They knew so little, it could be anything: the food, the habitat, an unseen pathogen.

Five days passed. Eve grew sicker by the hour until she lay in Honan's hand on her back with her legs sticking up, barely moving. Honan was frantic. He did not think she would survive the night. Desperate and acting on little more than a hunch, he prepared a concoction of glucose, calcium, water, and ground Melaleuca leaves. With the aid of a microscope and eyedropper, Honan dribbled the mixture into Eve's open mouth.

The effect was miraculous. Within the hour, Eve came back to life.

Of Fellenberg's 21 eggs, seven hatched: two females and five males. He was elated. The eggs can lay dormant for six to nine months. Once his eggs hatched, he closely observed them as the phasmids grew and passed into adulthood. He documented several molts along the way. Juveniles are bright green, and they are diurnal, meaning they are awake during the day. With each molt, their exoskeletons grow darker, and they slowly transition into nocturnal creatures. By the time they are adults, they are black.

Seven of Eve's nine eggs hatched.[9] Once recovered from her mysterious ordeal, Eve went on to live another 15 months; Adam lived another 18 months. Seventy-two of Eve's total 248 eggs produced offspring, but many of these died. Only 24 made it to adulthood. Honan noticed with alarm that the eggs of future generations were becoming smaller and more fragile. He worried about the lack of genetic diversity. That's when Fellenberg came to the rescue. He sent Honan some of his male phasmids to improve the gene pool. Within 12 months, Honan's Melbourne population was about 600 strong.

Captive populations around the world were quickly established as a security measure in Toronto, Canada; Bristol, England; and San Diego, United States. Today, a captive population also lives on Lord Howe Island.

It wasn't until 2017 that technology became advanced enough to run the DNA test to compare the old museum specimens from Lord Howe Island against the Ball's Pyramid phasmids. Guess

---

[9] They hatched on September 7, 2003, Australia's Threatened Species Day. It is the day remembered for the death of the last Tasmanian tiger (also known as the thylacine) in 1936.

what? They were a match. There was less than 1% difference in their DNA, easily within the margin of error.

This made everyone giddy, and you can see why. These findings greatly increase the likelihood that the captively bred phasmids could successfully be reintroduced to Lord Howe Island.

THE RAT PROBLEM

Not so fast, you say. *What about the rats? They're still a threat, aren't they?* They are, indeed. Which is why we can't run our VIP[10] victory lap just yet. There's one last step to take to make the island safe for the phasmids to return. The rats must go.

Go? Go where?

Brace yourself for this painful truth: We have to kill the rats. All of them.

Whaaaaaaat?! Kill an ENTIRE ISLAND FULL OF RATS?

Yes.

I know. It's kind of upsetting. Hear me out. These two species cannot coexist on Lord Howe Island. We know that. The phasmids, well, they were there first. More importantly, the island *needs* the phasmids.

Here's why: Funny thing about when something goes missing. Especially when something small and unassuming goes missing. Nothing happens at first, but over time, their absence starts to have consequences.

Dr. Carlile helps put this into perspective for us. He says that the Lord Howe Island stick insects were the cows of the island, the primary grazers of vegetation there. They were a fundamental part to the nutrient cycling because they were breaking down the

---

10 Very Important Phasmid

cellulose, then pooping it out as frass, which could be more easily incorporated into the soil. As a functioning part of the biodiversity of the island, they were critical.

They weren't alone in the task. There was also the wood-eating cockroach (another casualty of the rats). The cockroaches played a big role by dragging leaves underground and breaking them up. The seabirds ate both the cockroaches and the phasmids—in moderation, of course—which super-charged *their* poo into a nutrient-rich cocktail that allowed the soil to be conditioned, which in turn supported the growth of vibrant forests.

The forests, certainly in the last 100 years, have been missing a lot of inputs that would allow them to retain their vigor. Without those inputs, the forests become vulnerable. Windstorms sweep across the island and tear up all the vegetation, and it's hard for the island to recover. You end up with more vine growth instead of strong forest growth.

Factor in, too, all the inputs from people coming to the island over time—because there were no indigenous people here. People brought their pigs and goats and dogs, not to mention their weeds and other plants that don't belong, and before you know it, the once-vibrant, functioning ecosystem was severely hampered.

So the rats can't stay if we are to fix things. Don't think for a second that the rats should be left alone because what's done is done to the phasmids. There are other species at risk. Rentz's strong stick insect (*Davidrentzia valida*)—discovered in 1986 and only glimpsed a dozen times since—could be the next rat target. The little mountain palm. It's now listed as critically endangered because the dang rats scarf down its seeds and nibble its shoots.

Don't get me started. I could go on all day. Just trust me on this. The rats have to go. Period.

So how does one go about a task like this? It's not easy. Look at you . . . You didn't know anything about this when you started reading this chapter, and I can tell you're feeling bad for the rats.

The whole notion is so controversial that it's taken more than a decade of talking about it to finally convince the 350 or so inhabitants on Lord Howe that this is the way to go. In 2012, the Australian government and New South Wales government finally gave money for a project to eradicate the rats. Fortunately, both Australia and New Zealand have had successful eradication programs in the past, and we can learn from them. Lord Howe Island will be the largest inhabited island to be cleared of pests.

I'm not going to lie: Logistically, it's a nightmare. This project is underway at the time of this writing. Here's what is happening:

In the summer of 2019, the unpopulated areas of the island were dosed with 46 tons of rat poison. GPS-guided helicopters dropped pellets laced with brodifacoum, an anticoagulant.[11] From May through October, a 50-person crew set bait stations on the ground in areas where people live. Of course, before any of that happened, great pains were taken to capture certain species such as the endangered wood hen and the currawong.[12] Monitoring and captive management began in April 2019 and continues into the fall of 2021.

If everything goes according to plan, the island will become

---

[11] It basically prevents the rat's body from recycling vitamin K, which is needed to clot blood. Once animals run out of vitamin K, they can bleed to death.

[12] If you know what a currawong is, my hat is off to you. I had to look it up. It's a type of bird. Charmingly, it is known for its penchant to scavenge poisoned rat carcasses.

rat-free. To be certain they are all gone, the island must be monitored for two years. Two permanent biosecurity dogs and their handlers will monitor rodent activity. If two years pass with no signs of rats, then and only then could conservationists establish a small community of phasmids. Ideally, they could start on Blackburn Island first, about 700 meters to the west of Lord Howe Island, and work from there.

Then we're done, right? Sort of. In the right conditions, the phasmids could go wild. They might eat everything on the island, and then you have another sort of problem. Conservationists would probably reintroduce a native owl species to help keep the stick insects in check. I could tell you what happened to *those* originally, but we'd be here all day. Suffice it to say, when all is said and done, saving the phasmids and reintroducing them to their island home will impact no fewer than 70 species of plants and animals and will take decades more to accomplish.

It's doable. And noble. And a worthy project.

We tend to think these sorts of things are rare and happen in faraway places. Losing the phasmid isn't rare. Not anymore. It's happening a lot, and it's happening all over the world. We're ridiculously lucky with the phasmids. In most cases, we don't get a second chance to rescue the things we've lost.

By the time you've finished reading this book, I'm hoping you'll see this for yourself. I'm also hoping you will feel more connected to our world and, yes, even to our tiny insect friends. There are things you and I can do to help for sure, but I want you to know that the insects themselves are doing their part to survive. In the next chapter, I'll show you how amazingly resilient they can be.

9.

REPRODUCTION AND RESILIENCE

I KNEW THIS would be a tough sell after the ant incident. And the ants, well, they're so much smaller and more benign. This was much bigger and much more unnerving. This was cockroaches. Cockroaches are not something you invite into your home. They are something you fear, and if you are so unfortunate to find them in your home, you set that part of your house on fire and sleep with a can of Raid under your pillow just in case some of them didn't burn.

My battle cry of "In the name of Science!" fell on deaf ears with Chuck, Liam, and Devin. They weren't interested in any more science from me. I tried to explain that these were Madagascar hissing cockroaches[1]—a totally different creature than, say, a cockroach we might find behind a dumpster in New York City.

"These are bigger, right?" Chuck asked.

"They are, yes. But in a way, that's very helpful."

"How so?"

"They're easier to observe that way. You can see their anatomy

---

[1] *Gromphadorhina portentosa*

more clearly." *You can also see them coming if they were to accidentally escape*, I thought. Which they would not.

I noticed Chuck protectively crossing his arms over his chest.

"And you want to bring them into our home. And keep them as pets?" he said.

"One. I want to bring *one* into our home. In a small habitat with a secure lid. Not really as a pet, exactly. Just, you know, for observational purposes. They are amazing scavengers."

"And how long does a Madagascar hissing cockroach live, exactly?"

"I'll keep it in the sunroom," I answered, completely avoiding this question but preemptively answering the question I knew he was going to ask next. He didn't need to know *right now* that they can live for five years.

"It will be safe out there. You guys won't even know it's in the house," I said brightly, smiling at my less-than-happy family.

Liam shook his head. "I don't know, Mom . . . You're on your own with this one. They sound really gross."

"They aren't gross, Liam! I mean, they *are*. They're definitely gross. They're huge . . . and . . . and . . . cockroachy. But they aren't *nasty* like people think. They are actually quite fastidious."

The faces looking back at me were unilaterally unimpressed by the alleged fastidiousness of the cockroach. Devin looked up from his iPad, took one headphone off his ear, and said, "What are we talking about?"

"Never mind," I said.

They'll come around, I thought to myself later, as I placed the order.

The box arrived on Earth Day. The FedEx guy raised an

eyebrow as he set it down on the porch. Can't say I blame him. The box was plastered with bright red warnings: LIVE MATERIAL. AVOID HEAT AND FROST. PERISHABLE. I walked the box quickly through the house to avoid any family members and went straight to the sunroom.

My heart was beating a little fast. I'm sure the cockroach was in some sort of container, right? It wasn't just running around free inside the box. Surely not. I carefully sliced the tape on the box and gingerly opened the flaps. The cockroach habitat was wrapped in bubble wrap. The cockroach was in a sealed carton like what take-out food arrives in. But my heart stopped anyway. The label said 2 HISSING COCKROACHES. Two? TWO. No, no, no. Not two. Just one. I only needed the one. Two . . . two could be a problem.

I read the instructions on how to set up the habitat—pretty easy, really. Line the bottom with wood chips . . . Add a branch for it (them?) to crawl on and a little place for it (them?) to hide . . . Wet a sponge for water to drink . . . Set out some food. They are scavengers, but they mostly like fruits and vegetables that have gone bad. And dog food. They like dog food. That wasn't a problem. Zorro could share some of his kibble. Done! Habitat complete. Now just add the cockroaches.

Right. The carton was taped shut pretty thoroughly, so I had to snip through all that first. I could definitely feel some movement inside. The adults don't have wings, so it wasn't like they could fly out at me when I lifted the lid. They're relatively slow-moving. So really, it's fine, it's fine, it's fine.

I pried off the lid, but all I saw were paper towels. I slid the mass of paper towels into the habitat, and there they were! One . . . two . . . THREE? Oh, crap. WHY DID THEY GIVE ME THREE? Do those people not know how to count? Were they having a SALE that I wasn't aware of? Buy one, get TWO FREE?

This was just the moment when Chuck walked into the sun-room.

If he wasn't keen on the presence of ONE cockroach in the house, you can imagine his delight at seeing three.

Once the lid was firmly in place, a loooooooong conversation ensued. Of course, he asked all the right and reasonable questions: Why do we have *three*? What are the *genders* of these three? What if they *mate*?

I barely had time to look at them, let alone determine whether they were male or female. Yes, it was possible they could mate. No, I didn't know the gestation; I needed to look that up. I read the instruction booklet and consulted Google: Males possess large horns on the pronutum (behind the head), while females have only small bumps. The antennae of males are hairy, while the antennae of females are relatively smooth. Finally, the behavior of males and females also differ: Only males are aggressive and only males hiss.

Okay. Fine. Easy enough to sort this out. I went back into the sunroom to take a better look.

Liam appeared. "Did your cockroaches come?"

"They did," I said. "You want to see?"

"Uhhhhhh," Liam said.

"I'll just show them to you real quick!"

The cockroaches appeared to have completely disappeared, but they had just burrowed under their wood chips a bit. I removed the very secure lid and moved aside a few wood chips so he could see them more clearly.

He looked down into the habitat.

"Uh . . . Mom . . . what is THAT?"

That's just never good. When someone says that to you in *that* way. It's just never good.

"What is what?" I said.

If ever there was a time to curse, it was now.

Something was coming out of the back end of one of the cockroaches. It was a long, white tubular thing of some sort. Liam and I looked at each other and our mouths dropped. We both knew what it was, even though we had no idea.

"Google it," I whispered.

He reached for his phone and spoke into it.

"What does a Madagascar hissing cockroach look like when it is giving birth?"

He turned the phone over to me. I looked at the picture. I looked in the habitat. I looked back at the picture.

No. This couldn't be. This couldn't happen. This cockroach was in my house for exactly 15 minutes and now it was giving BIRTH?

"Liam."

"Mom."

"Holy crap."

"I'd say so. How many eggs do you think are in that egg case?"

My hand went to my mouth to stifle an oncoming scream. I did know the answer to this question.

"There can be as many as 60," I said.

"Holy crap."

"Yeah. Don't tell Dad."

Liam lifted an eyebrow.

"Don't tell Dad . . . *yet*. I need time to think."

"What's to think about, Mom? This gross cockroach just gave birth. Dad's *going* to notice."

"It's an egg case. We might have some time before, you know, the nymphs come out."

"Nymphs?"

Chuck arrived.

Uh-oh. I was in trouble now.

SETTLING IN

The rest of the day passed uneasily. Based on my rudimentary cockroach knowledge, I determined we had one female (clearly) and two males. I tried to take some small comfort having read that sometimes females eject their egg cases if they aren't viable. I brought this up at breakfast the next morning. Liam dropped his spoon and looked at me devastated.

"You mean . . . she *miscarried*?" he whispered.

"She . . . she might have," I said hesitantly. Oh dear. Was that the wrong thing to say? "I'm not sure. I've never midwifed a cockroach before. It's possible."

"That is so . . . sad," he said.

"It can happen. You know, in transit. I mean, if the egg case isn't viable, then we won't be besieged with a lot of cockroach babies, so maybe that's okay."

Chuck just snorted.

Ken, Barbie, and Mitch settled in. Before you ask, yes, I named them . . . in the desperate hope that cute names would somehow deflect the fact that they were cockroaches living in our house on purpose. Their original habitat felt too small. It was built for one, not three and an egg case. So I moved everyone to larger accommodations—a 38-liter (10-gallon) fish tank. I gave them an empty egg carton to hide under, which they seemed to appreciate. They drank from the sponge. They nibbled on carrots, apples, and dog food. They plotted their escape. Cockroaches can climb anything. They scaled the glass walls with ease. Not a problem. At. All. The lid was on good and tight, but Chuck asked me to seal it with tape. Just in case.

I took them out on occasion to directly observe them. Once you touch them, they are hard to put down. Not because they are so irresistible. To be honest, they gave me the creeps. It's because their feet cling to you so securely, you can't seem to detach yourself from them. It can be a little panic-inducing if you are skeeved out by them already. I should emphasize for you that Madagascar hissing cockroaches are rather large. They're one of the largest cockroach species in the world. If I held my hand out flat, Barbie covered my palm from thumb to pinky. Barbie and Mitch seemed nice enough, but Ken was very grouchy. He hissed a lot. By and large, they seemed to be gentle, social creatures. Maybe it was okay to have three. Maybe they needed each other for company. I started to relax about the whole egg case thing.

Imagine my horror when, two days later, I checked on our dear friends and found a single baby cockroach sitting on top of the egg carton. Holy crap! Guess that meant the egg case was viable after all. Where there was one baby cockroach, there was liable to be . . . 51. Barbie gave birth to 51 babies. I nearly fainted.

That's the thing about insects: high fecundity and high fertility. Fecundity refers to the number of eggs a female can produce. Fertility refers to the number of eggs that hatch. When both are high and the gestation period is short—often as little as two to four weeks—you're talking about an organism with fairly high rates of reproductive success.

How did this even happen? I mean, I *know* how it happened. I'm not asking about *that*. I was wondering about that egg case. Madagascar hissing cockroaches are different from most cockroaches in that they're ovoviviparous. Ain't *that* a 10-dollar word! What it means is that the eggs hatch *inside* the body of the mom. Yeah. Most cockroaches deliver egg cases called oothecae. The female Madagascar hissing cockroach carries hers inside her body. She keeps it there for several weeks while the larvae develop. When they hatch, they exit the case by exiting her body. It looks like a live birth. Whatever you do, do *not* google this. Seriously. It's *eew gross*.

Okay. Fine. I'll wait.

Gross, right? I did warn you. Most of the time, those babies go shooting out of their momma, but sometimes, momma pushes the egg case out first, and the babies exit the egg case outside of her body. Which must be what happened in Barbie's case. The stress of traveling all the way to my house from wherever she came from may have triggered her to eject it. But it was clearly viable.

I was in over my head, and I knew it. I turned to Dr. Phil for help. Not *that* Dr. Phil. Dr. Phil Mulder is a professor and department head of the Department of Entomology and Plant Pathology at Oklahoma State University in Stillwater. Thanks to some heavy online searching, I discovered Dr. Mulder penned a helpful document entitled *Madagascar Hissing Cockroaches: Information and Care*. After reading it, I felt fairly certain Dr. Mulder would be able to help me. I sent him a panicked email. He confirmed my suspicions but inflamed my fears:

In light of the brood that was produced, you can probably get a sense that you don't want to get into the breeding cockroach business. It will become overwhelming, and eventually your living quarters will not smell very good.

Hmmmm . . . indeed. I now had a lot more mouths to feed. There was something else on my mind too: gestation. How long before Barbie got pregnant again? How long before her young were old enough to have babies of their own? Dr. Mulder said seven months. Madagascar cockroaches can live for five years under good conditions. And the conditions I made for them were very, very good.

EXTENDED FAMILIES

Madagascar cockroaches are no slouches when it comes to reproducing, but there are other insects that have them beat. The African driver ant (*Dorylus wilverthi*) can produce three to four million eggs *every 25 days*. While that's enough to make your head spin, that's only one queen in one colony.

Argentine ants live in colonies with multiple queens—as many as eight for every 1,000 workers. Each queen can lay up to 60 eggs per day. In a relatively short period, Argentine ants can create supercolonies. These megacolonies take over huge amounts of real estate. In Europe, one colony is believed to stretch for 6,000 kilometers (3,700 miles) along the Mediterranean coast. The California large colony in the US extends across 900 kilometers (560 miles) along the coast of California.

Cabbage aphids (*Brevicoryne brassicae*) are another reproductively successful insect. In the spring, females turn out 5 to 10 genetic copies of themselves a day. A newly born summer female is like a tiny nesting doll. She is born pregnant. Gestation in an aphid, under the right conditions, is a week or maybe less. As temperatures warm, some of the eggs hatch into males. When the males are old enough, they mate with females, and the whole thing starts over again. The average life span of an aphid is about a month. Given all these factors, it is conceivable that a single aphid could generate enough descendants to cover the planet in a layer of aphids 149 kilometers (92.5 miles) deep *within a year.*[2]

This struck fear in my heart. I don't have an aphid problem, but I had the makings of a serious cockroach problem. If we assumed that half of Barbie's 51 babies are female and gestation for a Madagascar cockroach is about 60 days, just how quickly and how extensively can this intrusion—that's what a group of cockroaches is called—grow? We also couldn't discount Barbie, who could give birth three more times and be halfway through her fifth pregnancy while her first batch of babies matures.

DOING THE MATH

I needed to know what I was up against, but this was too much math for me to work out on my own. I decided to approach Mr. Coster. Mr. Coster is one of Wilson High School's finest math teachers. Liam had him for Precalculus and Devin had him for Algebra I. I wondered if he would think my request for help was strange, but Mr. Coster didn't seem fazed by my story at all.

---

[2] Shocking! But calm down. In reality, mortality is high for these guys because they have a lot of predators.

Which probably should've been troubling to us both. Instead, he seemed almost giddy at the prospect of trying to solve what he called my "exponential growth problem." He asked me a couple of clarifying questions. For the sake of the math, we assumed half of the babies would always be female, and each female would have the midpoint number of babies (40).

A few days later, Mr. Coster came back with a spreadsheet and a few caveats. First, he projected the data out for five years—Barbie's potential life span. But he wanted me to understand that what he created was a highly idealized model in which there were no predators, 100% of the babies reached maturity, and no one ever died. It also assumed an infinite amount of resources and an environment that supported this growing population.

Got it. I understood we were just running the numbers here. Hit me. What was the number?

"112.2 billion cockroaches."

My knees buckled a little. I checked the spreadsheet. It was pages of math with annual summaries:

After one year: *2,640 cockroaches*
After two years: *189,280 cockroaches*
After three years: *25.12 million cockroaches*
After four years: *1.78 billion cockroaches*
After five years: *112.2 billion cockroaches*

Holy crap.

"What are you going to do?" he asked me.

No. Idea.

Meanwhile, the cockroach situation was becoming a situation for other reasons. The babies kept escaping. *Didn't you say you taped the lid down?* Yes, yes, yes, I did. I ran packing tape all along the entire lid. Nothing should have been able to escape. But somehow they broke free. The tape wasn't even slowing them down. Like little insectoid ninjas, they were able to slip past its stickiness without so much as losing a leg. It happened five or six times. Luckily, I spotted the odd, errant baby making its way out of the sunroom before anyone else had. I was always able to get them back into the habitat, but it was getting to be a full-time job. Living with cockroaches could not be a long-term situation.

That said, I couldn't just release them to the wild. I needed to find a home for them—and by home I really meant a secure facility that was not my own home. Dr. Mulder suggested I contact the Smithsonian. They have an insect zoo and might be willing to take the whole clan off my hands. Sounds great! Except, no takers at the Smithsonian. I tried half a dozen different institutions and departments of higher learning but came up empty. I had one last thing to try: I reached out to my good friend, Paula Goldberg.

Paula sits on the board of City Wildlife, the wildlife rescue and rehabilitation outfit in Washington, DC, that I mentioned to you earlier. Paula taught me everything I know about wildlife rescue. In my volunteer time at City Wildlife, I've helped tend to many sick, injured, or orphaned birds, squirrels, turtles, opossums, the odd woodchuck, you name it.

I wrote her a quick email describing the circumstances:

Paula,

Have landed myself in a bit of a situation—but all in the name of SCIENCE—and could now use some sage advice.

With the insect book well underway, I acquired several hissing cockroaches from the Carolina Biological Supply Company. I had hoped to observe ONE and was prepared to keep it for its life span. I thought I had ordered one, but three came. Within 15 minutes of arriving at my house, Barbie ejected an egg case and 51 babies appeared.

If you do the math . . . well, it's terrifying how many cockroaches I will soon have on my hands. I know it is not wise to release them, so I have no plans to do that. But I would feel rather badly about sticking them in the freezer for a speedy death. I don't suppose you know of anyone who has a need for them? It's a rather strange problem to have.

Within minutes, Paula responded:

Can you call me? I'm laughing too hard, and I can't type well as it is.

Paula and I had a long conversation, but it landed on the circle of life. It seemed there were some hungry ducks at one of the wildlife rehab centers, and we agreed that becoming another animal's food source was better than dying a pointless death in the freezer. I felt a little bit bad about this decision, but having worked in rescue operations before, I also recognized the need for live food in some recovering animals. So . . . I surrendered the entire tank into Paula's most capable hands. The ducks, she reported back later, found the cockroaches delicious.

The lesson we are to learn from this experience, my friends, is never try to host cockroaches. It can quickly get out of hand. Of course, with all the excitement over the cockroaches, no one in the family even noticed me smuggling in the little praying mantis habitat. The one I hung in the dining room window that was holding a praying mantis egg case. The egg case that was probably housing 150 baby praying mantises.

MOTHERHOOD, PRAYING-MANTIS STYLE

I don't know if you've ever seen a praying mantis egg case before, but they are a wonder. The mother praying mantis lays her eggs and then covers them with a frothy substance. It quickly hardens into a light and foamy structure that's like Styrofoam. Now before you judge me and question my decision-making capabilities, let me just defend myself and say that harboring a praying mantis egg case in your dining room, while strange, is a perfectly okay thing to do. Because after several months, the eggs will hatch, and you will have tons of little mantises running around, and you can release them into your yard, and they will gobble up mosquitoes. No lifelong commitment to them or anything. At least, that's what I hoped would happen to mine.

Of course, first there was that part about the flies . . . In preparation for the eggs to hatch, it's recommended that you purchase a tube of fruit flies (*Drosophila melanogaster*). The flies serve as an immediate food source for your mantises. They will be ravenous when they hatch, and praying mantises only eat live prey. My flies didn't look very happy about being stuffed in a little tube. I'm sure they would've been even less happy to know my intentions for them. They glared at me with their tiny red eyes.

The fly tube came with a second tube filled with a sticky paste. It was labeled as food. According to the instructions, the flies were to live in their tube for the first week. At the start of the second week, I was to open the praying mantis habitat where the egg case was resting, toss in the open food tub, uncork the fly tube, and toss that in too—being quick to zip the whole sucker up as fast as humanly possible and walk away.

Somehow I managed to do this without releasing any angry flies by mistake. Now I was supposed to wait. For weeks I didn't know exactly when (or frankly IF) the praying mantises would actually hatch. So until they did, I had this fly-infested habitat hanging in my dining room.

It didn't take long for the family to notice, but oddly, no one really said anything—maybe because they were all suffering from cockroach trauma. But every day, I dutifully stood on a chair to peek inside to see if anything was happening. It was a whole lot of nothing until one glorious day. I climbed up and the inside of the habitat was teeming with teeny, tiny praying mantises. Even better, I saw babies still pouring out of the egg case. I was screaming like a maniac for my family to come as I watched these things squeeeeeeeeze themselves out of tiny holes in the egg case. Easily more than 100 came spilling out. They were so, so small. I felt absolutely elated. If you could have seen and felt that egg case—you just cannot believe that anything alive could come from it. But here they all were, streaming out! Even the boys, now hardened by several insect experiences, couldn't help but marvel. Nature, man! Nature is freaking *awesome*.

Our new friends made quick work of the flies. *Nom, nom, nom.* I waited only a few days for the babies to get a little bigger and a little stronger before releasing them to the backyard. They scattered to the four winds, searching for their next meal. Gone in an instant after such a long wait. But so worth it. So worth seeing their amazing emergence.

About four months later, I was working at my desk when I spotted something insect-like on the window. Instantly, I knew what I was looking at. My heart leaped. I ran outside. What had been pale and smaller than my pinkie fingernail was now dark green and brown and as long as my hand. It was absolutely magnificent.

The praying mantis stayed very still while I gawked at it and took pictures. I dragged the entire family outside to see it. Was it one of our babies? I couldn't say for sure, but in 13 years of living in this house, I'd never seen a praying mantis in our yard. Not once. I'd like to say it was one of ours. Healthy and thriving. To me, it felt nothing short of miraculous. Was it the only one that made it to adulthood

out of all that hatched? I don't know. All I can tell you is, to see it after so many months was a gift. Thanks, Universe.

Earlier in this book, we talked about how successful insects are. We pointed to a number of physical characteristics that contribute to this success. Their small size, for example. Their tough exo-skeletons. Their bright warning colors. Their ability to fly and escape predators or to relocate to places where resources might be more plentiful. And, as you can now see, insects are highly suc-cessful organisms reproductively speaking too. Still another factor that contributes to their success is their resiliency.

STAYING POWER

It can't be backed up by hard science, but I secretly wonder if some insects earn their resilience badge just because we are too afraid of them to approach them and try to kill them. Looking at you, cave crickets. You know these guys, right? From the family Rhaphido-phoridae. They go by many names: cave crickets, camel crickets, cave wētās, spider crickets,[3] or even land shrimp.

They look like a cross between a spider and a steroidal cricket. They have massive, drumstick-shaped thighs and curved, hump-backed spines. You'll find these wingless, nocturnal creatures in basements or other dark, dank places. They leap when they are frightened because it's the only defense mechanism they've got. Instead of hopping *away* from you, though, they hop *toward* you. It's terrifying.

There is something so heinous about the cave cricket that, when you encounter one, you seek only its destruction. Good luck to you on that. Experience tells me they are indestructible. You can throw things at them . . . try to step on them—it doesn't matter.

---

[3] Sometimes shortened to criders or sprickets

You can unleash an entire can of Raid on a single creature only to find it still moving afterward, prompting you to scream, "Kill it! Kill it with fire!"[4] It makes no difference, though. That thing will keep coming at you.

We know—of course we know—that cave crickets cannot harm us in any way. They cannot bite us or sting us. They cannot even fly. All they can do is lurch forward while dripping in toxic foam, backing us into a corner. By then, we are cowering and crying for our mommas, too afraid of them to strike the killing blow. And that, my friend, is how they stay alive!

NUCLEAR COCKROACH?

Many insects are resilient. Everyone always jokes about cockroaches, right? That cockroaches will be the last things standing after we obliterate ourselves in a nuclear disaster or some such? Ha ha. I'll be honest with you. The survivability of the German cockroach (*Blattella germanica*) is no joke. Could it *literally* survive a nuclear blast? The answer is yes but . . .

It can. But its quality of life will be poor. How do we know this? In the aftermath of the atomic bombs dropped on Nagasaki and Hiroshima, Japan, in 1945, there was anecdotal evidence that cockroaches had been spotted alive among the rubble. Earlier insect studies indicated that such a thing was possible. In 1919, W. P. Davey was experimenting with radiation when he hosed down a number of unsuspecting flour beetles with small doses of X-rays. Fully expecting the X-rays to outright kill the little buggers, you can imagine his surprise when the X-rays actually caused them to live longer. *Well, that can't be right,* thought another researcher, J. M. Cork. He

---

[4] I'm not saying this ever happened to me, exactly. But I'm not *not* saying it either.

repeated the experiment in 1957 . . . with the same results. That's not to say that exposing living things to radiation is necessarily a good thing, though. H. J. Muller's experiments in 1927 revealed that X-rays could cause mutations in fruit flies. In 1959, doctors Wharton and Wharton asked the question: Can radiation interfere with cockroach fertility? (Yes.) So that was interesting.

It wasn't until 2008 that anyone really took on the atomic bomb cockroach theory. You know the Discovery Channel's *MythBusters* show? The *MythBusters* team tested cockroaches, flour beetles, and fruit flies against radioactive material to determine their tolerance.

The insects were exposed to three levels of radiation from cobalt-60 for a month: 1,000 rads; 10,000 rads; and 100,000 rads. For context, the gamma rays released by the Hiroshima bomb were about 10,000 rads.[5] After 30 days, half of the cockroaches exposed to 1,000 rads were still alive; about 30% of the cockroaches in the 10,000 rad group were alive; and none of the insects in the 100,000 rad group survived. That's not a surprise. It's highly unlikely that anything standing at ground zero of a nuclear blast is going to make it. Temperatures alone at the epicenter of the blast reach about 10,000,000°C (18,000,000°F). If you're outside that initial blast area, you may have a fighting chance of surviving, but exposure to the blast and subsequent fallout will most certainly do damage to you.

It was calculated that cockroaches were able to survive radiation doses 10 times higher than those that would be lethal to humans.[6]

[5] The radiation-absorbed dose (rad) is the amount of energy from any type of ionizing radiation deposited in any medium (e.g., water, tissue, air). An absorbed dose of 1 rad means that 1 gram of material absorbed 100 ergs of energy (a small but measurable amount) as a result of exposure to radiation.
[6] Both the fruit flies and the flour beetles fared even better than the cockroaches, if you can imagine.

How? Researchers think the most plausible explanation is that the cell cycles of cockroaches are slower than those of people. As it happens, cells are more susceptible to radiation damage when they are dividing. Peoples' cells are constantly dividing, but cockroach cells usually divide right before they molt. Depending on when a blast happens, only some of the cockroaches would be vulnerable because of cell division triggered by molting.

All nuking aside, cockroaches are, evolutionarily speaking, highly successful organisms. They've been around hundreds of million years—far longer than humans. They've outlived dinosaurs by 150 million years. They reproduce quickly and in large numbers. They also adapt to a wide variety of conditions and even develop tolerance and immunity to some toxic things. I mean, we are talking about an animal that can live several weeks without its own head. They aren't impervious to everything, but they're pretty darn close.

NIGHTY NIGHT, BED BUGS

While we're on the subject of resilience, we should also probably talk about bed bugs. Bed bugs (*Cimex lectularius*) are what we like to call a nasty bit of business. That's not a technical designation, but I still say it's a fairly accurate one. They are pests to be sure— small, nocturnal creatures that feast on human blood. They have mouthparts that saw through human skin and inject it with anti-coagulants. As they suck your blood, their bite leaves behind a red and swollen mark that itches you later. Bed bugs start out the size of an apple seed but quickly bloat up as they consume three times their own weight in your blood during a single feeding. Having finished their meal, they drop away from their human host and do not need to feed again for a week or so.

Unfortunately, bed bugs have been a problem throughout human history. Even the ancient Egyptians, who were rather partial to insects, had a special spell designed to cast out bed bugs.[7] Alas, mixing wormwood and rue and chanting magical phrases proved to be no more successful than many of the other remedies people have tried—like smothering them with borax, dousing them with rubbing alcohol, or slathering them in tea tree oil. What about bleach? Doesn't bleach kill everything? No, bed bugs are extremely hard to get rid of and most successful attempts are the result of a complex strategy that involves heat-treating your living space and engaging in a whole lot of vacuuming.

What makes bed bugs so invincible? They're small, flat, and wildly adept at squeezing themselves into tiny spaces. This makes them hard to find when you're trying to hunt them down. If resources are scarce, bed bugs can survive a shockingly long time without eating; researchers have documented stretches as long as 550 days. Under the right conditions, it doesn't take long for one bed bug to become a full-blown infestation. A single pregnant female can lay up to 500 eggs during her year-long lifetime. Her brood can grow exponentially; three to four generations could be produced in one year.[8]

For these reasons, bed bugs have been the bane of our existence for a really long time despite our ardent efforts to eradicate them. We nearly succeeded in the 1940s after experimenting with DDT. Dichloro-diphenyl-trichloroethane was one of the most effective

---

[7] Interestingly, to cure snakebite, ancient Egyptians would drink them. Indeed, throughout history, well-intentioned doctors created a host of bed bug–based cures for all manner of illnesses and conditions, including jaundice, vomiting, lethargy, and ear infections. The 18th-century French naturalist Jean-Étienne Guettard recommended eating bed bugs to treat hysteria. I could see how such a thing might *induce* hysteria.
[8] Bed bugs reproduce most quickly in 70–82°F, which happens to be the range where most people keep their thermostats.

pesticides ever created for killing insects. It was colorless, tasteless, nearly odorless, and cheap to produce, and it wreaked absolute havoc on an insect's nervous system. DDT interferes with normal nerve impulses, leading to tremors, convulsions, and eventually death. While our skin acted as a barrier to DDT, insect exoskeletons quickly absorbed it. Here was the answer to all our insect woes. No more bed bugs! Its liberal use decimated bed bug populations around the world.

DDT had been initially developed to combat malaria, typhus, and the other insect-borne human diseases. It was also used to control insect problems in crop and livestock production. DDT, however, can be absorbed by eating, breathing, or touching products contaminated with it. While exposure to it (largely through food) didn't outright kill people, it could make them sick: nausea, dizziness, confusion, headache, lethargy, vomiting, and tremors. And it affected other animals, like birds, in terrible ways. It caused eggshell thinning and embryo deaths. It was highly toxic to fish and amphibians. We came to realize that DDT can fully saturate an environment, and its residue remains long after it's been applied—as long as 15 years under some conditions.

Of course, this residue was what proved to be the bed bugs' undoing. Bed bugs stay hidden in cracks during the day when you're not around to feed on. Earlier spray deterrents often dissipated quickly or never reached the areas where bed bugs were hanging out, but because DDT left a residue, the insects would walk through it whenever they surfaced. Between the 1960s and the 1990s, bed bug infestations were all but unheard of in the United States. Eventually that changed.

While some countries continue to use it, the United States banned DDT in 1972. Even before the ban, bed bugs were starting to become resistant to it. Pockets of resistant bed bugs were evolving around the world. Then in the 1980s and 1990s, the US saw a shift in domestic and international travel. Through airline deregulation and new connections being made with other countries, travel got a lot easier and cheaper. All this travel may have helped spread the resistant bed bugs.

The resurgence was shocking. In New York alone, bed bug infestations increased by 2,000% between 2004 and 2009.[9] These were not your granddad's bed bugs. For starters, they evolved to have thicker exoskeletons—about 15% thicker. This evolution made it less likely that insecticides could infiltrate their shells. That's not all. Today's bed bugs have what's called a knockdown resistance. They produce enzymes that break down older insecticides, rendering them less toxic to the bed bugs. Getting rid of them now will be so much harder than it was the first time.

FIRE ANT GROUPTHINK

Up until this point, we've been talking about how insects survive despite individual threats like Raid-carrying stalkers or DDT-wielding exterminators. Now we want to turn our attention to insects that survive threats as a collective—sort of like the fictional Borg, if you ever watched *Star Trek: The Next Generation*. The Borg[10] was a cybernetic organism, its parts linked together in a hive mind. It moved about the universe, co-opting the best parts of other alien species by injecting individuals with nanoprobes and surgically augmenting them with creepy-looking electronics in a

---

[9] Lest you think this can't be serious, in one case, a 60-year-old man had to be hospitalized due to blood loss brought on by feeding bed bugs.
[10] "Resistance is futile." —the Borg

quest to achieve perfection. In behavior, fire ants (from the genus *Solenopsis*) are kind of the insect version of the Borg.

Ants are social insects. Social insects live in colonies that have three defining characteristics. They work as an integrated group. They divide all labor. And they have overlapping generations with young and old individuals living together. Ants use a collective intelligence to solve complex problems such as finding the closest food source or creating complex tunnel systems. Each singular ant behaves like a neuron in a central brain. They interact and react to the behavior of their closest neighbor.

This provides the nest with a certain amount of resilience when faced with disruptions. If disaster strikes or if members of the colony are lost, other ants can switch jobs to continue the work. By working together, a small glitch can be fixed before it becomes a system-wide failure.

Fire ants live together in underground colonies that can number in the hundreds of thousands. They are native to Brazil, where they face periodic floods. While other underground insects may drown during severe flooding, fire ants work in a specific way to ensure their survival. When flooding occurs, the ants seek higher ground and begin forming rafts by linking legs. The waxy covering that coats their exoskeletons (combined with small, bubble-holding hairs) help keep the ants waterproof. Ants in the water on the bottom of the raft constantly circulate to the top to get air. All life stages of the colony—from queen to eggs—exist on this mat of floating ants. Some workers even carry eggs in their mouths to keep them safe.

The raft can stay afloat for days. When it hits land, the ants keep building. They make living bridges that allow the entire colony to scramble to safety. Afterward, using their bodies, they construct temporary aboveground structures to provide shelter for the few days it takes to re-dig underground tunnels. All the while, the ants that form the temporary shelter are continuously moving but still preserving the structure. It is this living architecture that researchers are so interested in.

David Hu and Nathan Mlot from Georgia Tech are taking a hard look at how red imported fire ants[11] construct bridges, rafts, and shelters using their own bodies as building material. The ants continuously reorganize their structures to maintain stability, and their movement has a fluid, liquid quality to it. The ants actively change their structure to accommodate a stress (like a rock in a stream) but then bounce back into place afterward. Can what we know about fire ants be useful to us in terms of how we build or how we think? It's possible.

Whether we are talking about cockroaches, bed bugs, fire ants, or other insects, you can see that they are highly successful organisms despite our constant human attempts to wipe them out or control them. That gives us an odd comfort, I suppose. I'd rather not leave the world in the hands of the cockroaches. Oh, I know they'll do their part to survive. But the thing that I'm constantly wondering about is this: Can we do our part to keep insects—not just the ugly and reviled ones—alive? I wholeheartedly believe the answer to that question is yes. I think, just by the sheer fact that you are still with me and still reading this, that you must believe so too. Will it be hard? Yes and no. There's a lot we can do on our own and together. Let's head into Chapter Ten—our final chapter—to see what we can do.

---

[11] These ants were accidentally introduced into the United States in the 1930s and have been causing trouble ever since.

10.

DOING OUR PART

I COULD STAND up on my chair and scream to you *THE SKY IS FALLING*, and I wouldn't be wrong. That might get your attention for a few minutes. But quite honestly, the insect problem is much more nuanced than that, and the solutions are going to require more than a minute of your time.

I'm not going to sugarcoat this: We've got some challenges. First of all, so few insects have been assessed for their Red List status, we don't even know what our true insect universe is. It's highly probable that we are losing insect species to extinction on a rolling basis and losing things we never knew we had. Second, of the things that we know are in trouble, most are living in unprotected habitats. Their fates are really left up to Fate. Third, the traditional approach to animal conservation involving restrictions on hunting and wildlife trade can't really help us here, considering that most insects aren't hunted.[1] No, in fact, we've got a bit of the opposite problem: Most people want to avoid insects.

You can't employ the same conservation strategies for all animals. Something large and charismatic—like the panda bear—has different needs from something small and less appealing—like the

---

[1] Some are, to be sure, but the bulk of them are not.

Zayante band-winged grasshopper (*Trimerotropis infantilis*) (look it up). We are hardwired to like animals that are similar to us either physically or behaviorally, or, frankly, to like animals that are cute. The more an animal meets these criteria, the more likely we are to feel a moral duty to care for it. The fact that chimpanzees use tools connects them to us. They use tools; we use tools. We are alike in that way. The fact that house flies vomit on their food before they eat it? Not so much.

Well, foo. We can't all be pandas, can we? No. So what do we do? For starters, we can look to the world's governments for some help. So much of the research that we talked about in Chapter Seven has registered, and nations are beginning to take action. The 2018 Agriculture Improvement Act is providing millions of dollars for pollinator research and conservation across the United States. The European Union endorsed a major initiative that same year to protect pollinators. Germany followed in 2019 by pledging funding for insect conservation, monitoring, and research. The Swedish government plans to spend $25 million on pollinator protection initiatives over the next three years. In 2020, the government of Costa Rica endorsed an effort worth $100 million to inventory and sequence every multicellular creature in the country over a decade, much of which will be insects.

Further, the European Union and several other countries have passed legislation to restrict use of some pesticides. Several neonicotinoids have been banned from use on field crops. The Environmental Protection Agency released an interim decision in 2020 on the use of neonicotinoids in the US aimed at protecting pollinators.

There's more that can be done on a global scale. I actually made a list! These are fairly general suggestions, but think of where we'd be if countries actively pursued these goals:

1. Encourage more people to study insects and become experts so we can increase the number of people identifying and classifying species.
2. Revive, support, or create more insect-monitoring programs so we can continue to build databases of what we have and what's in trouble.
3. Increase research on what's driving extinctions.
4. Support innovative methods of pest control that don't rely on chemicals.
5. Support and increase research on climate change and its impacts on insects.
6. Increase education and outreach efforts to improve public understanding of insects and their roles on Earth.

We can also—as governments, as communities, or as cities or towns—take direct actions in support of our buggy friends. Remember me telling you about light and noise pollution? As it happens, these are two of the easiest forms of pollution to deal with because once you remove them, they are gone. They don't linger in the environment like chemicals or plastics do. Simply turn off the lights and there! Done!

Maybe it's not *quite* so simple as that, but installing directional covers on outdoor lights so they only illuminate the areas where they are needed can cut down on light pollution. Making lights motion-activated so they're only on when people are around is another solution. Avoiding blue-white lights, which interfere with insects' daily rhythms—LED lights can be easily tuned to avoid harmful colors and flicker rates. These are actual, tangible steps that we can petition for in our local communities.

City planning authorities can also introduce policies to help reduce noise pollution. Transportation networks and construction timetables can be carefully managed. Low-noise road surfaces, low-noise tracks, redesigned street spaces, enforced speed limits, and reduced traffic volume—these all can help bring down the noise levels.

Or get this: Have your city plant some trees. Trees effectively muffle noise in urban settings, such as around major highways. Planting trees is a smart move, of course, because trees are aesthetically pleasing, improve air quality, and might become a food source or a home to insects. Efforts are already underway in Canada to reduce noise in another way. BC Ferries is working to reduce the effects of underwater noise along ferry routes that pass through critical habitats.

One thing about having a complex problem is there are myriad ways to attack it. We have agency. You and I can take actions as individuals to turn things around too. I'm always reminded of that snowflake parable:

"Tell me the weight of a snowflake," a sparrow asked a wild dove.

"Nothing more than nothing" was the answer.

(Yes, I know this is a book about insects. Just humor me.)

The sparrow told the dove a story. It was snowing, and the sparrow was sitting on a branch of a fir tree. For want of something better to do, the sparrow counted the snowflakes falling on a nearby branch.

"Their number was exactly 3,741,952," the sparrow said. "When the next snowflake dropped onto the branch—nothing more than nothing, as you say—the branch broke."

The sparrow flew away, leaving the dove to reflect on what it had said. *Perhaps,* the dove thought, *there is only one person's voice lacking for peace and justice to come about in the world.*

I always liked that story because it is a reminder to us that while we might feel like we are insignificant or our contributions might not matter, they actually do. Especially when they are pooled together with the efforts of others.

You and I have spent this entire book together, learning all the ways insects are vital to our planet. We've also learned about some of the threats they face. So . . . what can we *do* about it? You and me. Regular people.

You don't have to love insects to help them. If you still think they are *eeew gross* but understand how important they are, that's all I need. You've made a fairly big commitment already, you know. You read this book. And I thank you for that.

Unfortunately, I wasn't able to cover *everything* in this book. I never got around to telling you about the whole "milking cockroaches" movement.[2] Or the studies indicating bees can count. Or any of the stuff about how bees can be trained to sniff out explosives. Or that whole thing about how some sap-sucking insects can fling their own pee at an acceleration rate 20 times Earth's gravity—and no one knows why. We didn't delve into the

---

[2] Cockroach milk is a protein-rich, crystallized substance produced by *Diploptera punctata*. Some folks think it would be advantageous to harvest this milk, as it is a good source of protein, carbs, and fats. It is also considered a complete protein source, providing all nine essential amino acids. I cannot think of anything more repulsive, but there you go.

studies on whether insects have consciousness and ego. There's so much more we need to talk about. I just ran out of room and hit my deadline. You and I can keep talking, though.

At the end of this chapter, I made a list for you of some remarkable books that have helped me understand things better. If you can, look some of these up and give them a read. Then share what you know. Tell your parents, your friends, your neighbors. Tell total strangers. Help me spread the word. As you begin to make connections between insects and our world, you can help others to see them too.

Another thing we can both do is learn from experts. There are a couple of organizations that can get us up to speed on the latest research and the work being done. I know you (like me) like to know what's going on, so check out these organizations:

AMATEUR ENTOMOLOGISTS' SOCIETY

amentsoc.org

Founded in 1935, this organization is run by volunteers with an interest in entomology. Their goal is to promote the study of entomology, especially among amateurs and youth—both groups of which I think we fall into, right? The society produces a number of publications—newsletters, handbooks, and pamphlets with useful information—and they have a network of experts who can help with any questions you might have.

BUGLIFE

buglife.org.uk

Buglife is the only organization in Europe devoted to the conservation of all invertebrates. Their motto: "Saving the small things that run the planet." They do incredible work and have some really interesting resources.

ENTOMOLOGICAL SOCIETY OF AMERICA

entsoc.org

If you're really quite serious about insects, look into these folks. This is the largest organization in the world serving the professional and scientific needs of entomologists and individuals in related disciplines. Founded in 1889, ESA has more than 7,000 members affiliated with educational institutions, health agencies, private industry, and government. Members are researchers, teachers, extension service personnel, administrators, marketing representatives, research technicians, consultants, students, pest management professionals, and hobbyists.

XERCES SOCIETY FOR INVERTEBRATE CONSERVATION

xerces.org

This is an international nonprofit that protects the natural world through the conservation of invertebrates and their habitats. Their name (which is pronounced Zer-sees) comes from the now-extinct Xerces blue butterfly, which was the first butterfly known to go extinct in North America as a result of human activities. This organization is a science-based conservation organization. By utilizing applied research, engaging in advocacy, providing educational resources, addressing policy implications, and building community, they are hoping to make meaningful long-term conservation a reality.

These are just a handful of organizations, but you can also learn more about insects and conservation from any of the experts in this book. You can directly contact conservation groups that work in your area or cater to your favorite insect. You can go to museums or nature centers or your local library. There's a lot of information out there for us to tap into.

There's *stuff* we can *do* too, if you're interested. Some of you have gardens. Some of you don't. But all of us can plant stuff, even if it's a bucket of flowers put on our doorstep. Don't think for a second that a bucket of milkweed on your porch won't be greatly appreciated by a hungry monarch that needs to eat on the go. Don't have a doorstep? What about a windowsill box or rooftop? I don't know where you live, so I can't tell you exactly what to plant, but here are some things for you to think about:

PLANT NATIVE PLANTS

We'd all like to see our backyards abuzz with bees and bugs. That said, it's very important for you to either plant native plants or maintain the native plants you have. It's tempting to plop down something exotic, but try to resist the urge. Why?

Native plants evolved with native wildlife, so everything about native plants connects to native insects, birds, and other animals. The timing of when native plants bloom connects with the pollinator cycles in your area. Leaf growth of native plants corresponds with the feeding habits and reproductive cycles of native leaf eaters like caterpillars. As your insects find food and other resources, their populations grow.

Birds need native plants for the fibers and twigs to build their nests. Just as they are beginning to nest, the insect population is at its peak—so there are plenty of insects the birds can eat and feed to their chicks. Fruit, berry, and nut maturity are timed to bird and animal life cycles. Animals depend on specific plants to provide vital food for fall migration. Mammals rely on them to fatten up for hibernation. Everything is connected.

Do you know who is really knowledgeable about this? Dr. Douglas Tallamy. He's an entomologist who teaches at the University of Delaware. A few years back, he wrote a book called *Bringing Nature Home: How You Can Sustain Wildlife with Native Plants.* I urge you to read it because Dr. Tallamy explains some really key concepts: Native insects need native plants. Native birds and mammals need native plants *and* insects. What Dr. Tallamy tells us is this: By planting native plants, *you can have an immediate impact.* You can restore (or maintain) habitat for wildlife and preserve natural history. Wouldn't that be such a good feeling? To do something small that makes an immediate impact?

TIME YOUR PLANTINGS

This might take some planning and research on your part, but try planting things that flower at different times of the year. That way you'll attract more insect visitors—early spring, dead of summer, early fall. Your yard will look amazing, and you'll be the envy of your neighbors. They'll covet your yard. They'll think to themselves, *Why doesn't our yard look half as fabulous as their yard?* And because people are competitive by nature, this burning yard envy will spur them to action. Before you know it, your neighbors will be stepping up their game. They will plant more stuff. Better stuff. They will go all native too. See? By total accident, you will have transformed your block and made it an insect-friendly zone.

When you do this—notice I said *when* and not *if*—*when* you do this, you must let me know. You must send me pictures so I can bask in the glory of your yard, your neighbors' yards, and the great victory you have won for insects.

DON'T FORGET THE NIGHT FLIERS

We do like to support those daytime pollinators, don't we? Don't forget to show a little love to our night fliers too. Feed the moths! Plant jasmine, honeysuckle, or sweet rocket. Look for other nighttime bloomers that fit your area. You don't need to lose your mind over this. Don't redo your entire yard. Plant a few things in pots. Gussy up the space with some twinkling solar lights to attract the moths.[3] Done!

DON'T POISON THINGS

While we are speaking about what to plant and what not to plant, I must caution you about using chemicals. Many pesticides and insecticides are indiscriminate and end up killing a wide range of insects instead of just the ones we hope to target. Insects that are directly exposed to a pesticide can die, of course, but others can die when they eat contaminated nectar or prey. Little birdies and sweet hedgehogs can also be killed when they eat poisoned insects. You don't want to be responsible for dead birdies or hedgehogs, do you? Of course not.

Pesticides can also get into the roots of your plants and trees. They can get into your groundwater. If you're trying to get rid of a specific thing—like mosquitoes—there are other ways to do that.

---

[3] The local bats will probably thank you as well!

Instead of trying to go nuclear on your yard, why not try planting things that are naturally repellent to mosquitoes?

If you're looking for plants that smell great to us but not to mosquitoes, try citronella grass. Rosemary. Basil. Mint. Each of these are repulsive to mosquitoes but smell great to us and can also be used to add some zing to your cooking. If you don't mind making all the neighborhood cats loopy, you can also plant catnip. It's actually from the mint family and is an amazing mosquito repellent. One study found it to be 10 times more effective than DEET.

Lavender is another good choice. Mosquitoes hate it, but you know what loves it? Bees. It's popular with honey bees but even more so with bumbles. I think bumbles were built for lavender. A bumble bee needs only 1.1–1.4 seconds to extract nectar from a lavender flower. Honey bees have shorter tongues, so it takes them a bit longer to push their heads farther into the flower to reach the nectar. They take about 3.5 seconds per flower. A couple of seconds difference may not seem like much, but multiply that over several thousand flower visits, and it adds up.

Check out this math: On average, a lavender flower holds only 0.02 microliters of nectar. Given that a honey bee can hold 50 microliters in its honey stomach, it would need to visit about 2,500 lavender flowers before returning to the colony. At a rate of 3.5 seconds per flower, that's going to take somewhere in the neighborhood of two hours and 43 minutes for the honey bee to fill up, whereas the bumble can do it in a fraction of the time. I digress here a little . . . The point is you want to drive off mosquitoes. But in planting lavender, you're also luring a lot of bees. Go you!

You can multitask by planting marigolds too. Goodbye, mosquitoes, yes, but goodbye, aphids, thrips, whiteflies, Mexican bean beetles, squash bugs, and tomato hornworms too. These guys want to eat your vegetables and other plants but won't want to cross that border of stinky (to them) marigolds to do it. That's okay, you know. To want to drive off *some* insects.

We agree—do we not?—that some insects are pests. Yes, I think it's okay to say that. We don't really want to invite pests over. We only want to invite beneficial insects, right? And what might *those* be, you ask? Beneficial insects fall into three categories: pollinators, predators, and parasitizers.

We know who the pollinators are: bees, bees, and more bees—but also butterflies, moths, flies, even some beetles.[4] How about those predators? Do I really want predators in my yard? Sounds scary. Yes! You do want them, especially if they eliminate pests by eating them.

Ladybugs have such a sweet reputation, but they are fierce predators. They *hoover* aphids. A single ladybug can suck up 40 aphids an hour. Aphids are pests because they suck the nutrient-rich liquids out of plants, which weakens your plants and wreaks havoc on fruits and flowers.

Green lacewings feed on pollen and nectar as adults, but as larvae, they prey upon aphids too. You definitely want to welcome them to the yard. You have to be a little mindful with praying mantises. They have enormous appetites, eating aphids, leafhoppers, mosquitoes, caterpillars, and other soft-bodied insects when they are young. When they get older, they favor larger prey—beetles, grasshoppers, crickets, and other pest insects. They aren't too picky, and they may also gobble up a few of the friendly insects

---

[4] See Chapter Two if you need a refresher.

like butterflies. Heck, they've been known to eat *hummingbirds* before. They're pretty voracious.

Ground beetles are beneficial predators too. That's a whole category of beetles that eat a wide range of insects, including nematodes, caterpillars, thrips, weevils, slugs, and silverfish. Soldier beetles, assassin bugs, dragonflies, and robber flies are also good predators to have around.

Like predators, parasitizers also prey upon other insects but in a slightly different way. They lay their eggs on or in the bad bugs, and when the eggs hatch, the larvae feed on the host insects. Parasitic wasps are the main member of this category. They are very tiny, so you probably won't see them actively working to rid your yard of pests, but that's what they do.

While you're thinking about your outdoor space with an eye toward insects, you might also think about the ways that you can create special habitats within your space for them.

BEE A GOOD HOST

You may want to provide a hotel for insects to stay in. Do a little search for insect hotel or bee hotel, and you'll see a lot of attractive, clever little houses for insects to make their nests in. Most of them have little tubes or hidey-holes for the insects to inhabit.

I have to tell you this: We built some from scratch. Bee hotels. Of course we did. I recruited my parents, who are both skilled artists, and my dear friends from Glen Echo Pottery—Jan, Cathy, and Katy.

I described the plan: Dad would design the bee hotels, buy the lumber, cut it, and assemble the basic shapes. I would secure the necessary tubes—some cardboard, some bamboo—for the bees to lay their eggs. Mom would spearhead the decorating and finishing aspects. The potters would each take a hotel, finish it, and later display it in their yards or gardens to draw the bees.

Crowded around my dining room table one afternoon with our freshly built bee hotels, Mom laid out the ground rules for the team.

"No red," she told the group. "Brenna says that bees can't see red." That's true. "So, we'll have to use pukey pastels because that's what the bees seem to like." Also true. And true that my mom can't abide "pukey pastels" because she thinks soft colors are uninteresting. The ladies—Jan, Cathy, and Katy—didn't bat an eyelash. They went straight to work designing each hotel's look.

While we worked, I took this opportunity to ask my little group to name their favorite insects. All of us are nature lovers, so I figured this would be an easy question. Dead silence. Mom kept her head down, focused on her work. There was some clearing of throats.

Seriously, girls? Not one of you can think of one insect you *like*? C'mon. The default answer if you're really struggling would be bees, right? Since we are making BEE hotels?!

Cathy finally came up with a tentative answer: "Butterflies?"

"Oh, yes," Jan quickly seconded. "Butterflies. Those are my favorite too."

I looked to Katy.

"What are those praying mantis thingies that look like flowers?"

The orchid mantis (*Hymenopus coronatus*).

"Those are pretty cool."

I was struck with a funny realization. I had been on this insect kick for a while now. My friends here had eaten my cricket cookies. They had tasted the cricket brownies. They had listened to countless tales of insect adventures and mishaps, read pieces of this manuscript, and been subjected to close-up photos of some pretty disgusting-looking of bugs. And now they were painting bee hotels to install in their own yards. The reason they had put up with all that was because I had asked them to. How lucky am I to know such supportive and giving people? I know you have people in your life like this too. It's a really important thing to remember. Together we are always stronger. One bee hotel is great, but six bee hotels in six locations is even better, right? It might look like a small step, but to me if felt like a huge leap.

The team made quick work of the hotels. A lot of us went with stripes, painting each facet of the hotels in different colors. Mom painted colors on but then immediately wiped them off, giving her hotel a distressed look that definitely dampened the pukiness factor of those pastels. Like all great art projects, everyone had a different interpretation and result. They were pretty fancy hotels, I can tell you. When the paint was dry, we stuffed the hotels with 15-centimeter (6-inch) tubes. That spring, we set them in our yards and gardens and watched as mason bees and leafcutter bees moved into their new homes.

PILE UP DEAD STUFF

Let's face it—some of us lack the Martha Stewart gene and not all of us are as handy and skilled as my parents and pottery friends. Yet most of us are capable of making piles. I am especially gifted at making piles of books. My kids are pretty adept at forming piles of laundry. Whatever your pile-making specialty is, each of us can apply ourselves outside by piling up dead things. I'm talking sticks, dead logs, leaf debris, leftover bits from pruning—all that good stuff. Take that and pile it up in your yard somewhere. If you're artful about it, you can make it look like an intentional piece instead of the precursor to a bonfire. When your neighbors ask about it—and neighbors being what they are, they will ask about it—here's what I want you to tell them: *I am creating a habitat for insects that do a special job for Mother Earth.* Best to say it with your hands on your hips Wonder Woman–style so they know you aren't kidding around.

That's right, your little pile is dedicated to our decomposers who break that stuff down and enrich the soil. Solitary bees, beetles, and ground-dwelling insects thrive in those little maze gaps between rotting logs and flaking bark. See Chapter Two if you need a refresher on their capabilities.

Where you position this in your yard will determine, to a degree, what sorts of insects you might attract. If you put it somewhere shady, it's likely to rot more quickly, so you'll see fungi and mosses growing and beetles scuttling about. A sunny pile will dry and harden. Solitary bees and wasps that harvest wood for their nests will visit such a pile. You may not see a lot of the activity—a good deal of it will happen with your night-crew insects. Rest assured, though, your pile will get used.

While this may not look attractive, I am also going to make a tiny plea for deadheading and leaving little patches of your yard unmulched or unraked. If you can stomach it. If you leave deadheads until late spring, you can provide overwintering sites for a lot of insects. The same is true for leaving leaf piles in place. That's a big ask for some of us, I know. Many of us are neat freaks who want to sweep everything up to restore order to our little patch of the world. Yet if we can just hold off and do our cleanup chores in mid- to late spring, then we can really help our insect friends.

COMPOST YOUR WASTE

If possible, consider creating an outdoor compost heap. This isn't for everyone, but it can create a nice home for pillbugs, sowbugs, or woodlice. Several types of flies and some beetles (and the offspring of both) are sure to visit. Adult flies constantly scan the area and lay eggs on good larval food sources. Fruit fly larvae and black soldier flies are eager eaters of coffee grounds and kitchen waste. All these critters will feast on this rotting banquet you've put out for them. The vigorous tunneling action of hungry insects can help aerate the material. As they work their way through your stinking pile of goodness, they boost the presence of bacteria and fungi, which help break down the compost. Insects in and around your compost bin can be excellent food sources for hungry birds as well.

DIG A POND

If you do have space and inclination, think about building a pond. A pond is a different kind of insect habitat. If you build it, they will come. Ponds are quickly colonized by dragonflies, damselflies, pond skaters, backswimmers, water beetles, and aquatic larvae of many other insects.

DON'T FORGET THE DRINKS

Regardless of what you build or don't build, do give some thought to providing a water source, if possible. Insects need water. Herbivores mostly get water from the plants they eat. Honey bees need fresh water to drink. If you are a beekeeper in Washington, DC, you are required by law to provide a water source for your bees—it's that important.

The bees aren't just drinking it. They use water for cooling the hive by evaporation and for thinning honey to be fed to larvae. A strong hive on a hot day can use over 1 quart of water, which would take about 800 bees, each making about 50 trips to a water source. Carnivorous insects often have to get their water from a different source from their prey. Fill your birdbath, if you've got one. If not, leave a little saucer of water out for your insects on hot days. Change it frequently so it's fresh.

OPEN YOUR WALLET

Wait. Is all this too much nature for you? Too much being outdoors? Maybe you have no space of your own or gardening/ landscaping is not your thing. No problem. You don't have to get your hands dirty to help. You can help when you go shopping. I'm not kidding! Your purchasing power—what you buy and what you *don't* buy—can help insects.

When you wear cotton, make sure it's organic cotton. Non-organic cotton uses a lot of insecticides and pesticides to produce. Maybe you're not in the habit of looking at labels, but this is an easy thing to learn. The next time you are sizing things up at Urban Outfitters or wherever you go to look sharp, just take a little peek at that label. If you see nonorganic cotton, shop with your feet—walk on out of there and find something else. Something that works for Earth and our insect friends.

Do this with food too. You don't have to eat insects if you don't want to, but try to eat certified organic and non-GMO-verified foods. Genetically modified seeds are causing farmers to use more pesticides and herbicides, and we know that kills all kinds of insects. Whenever possible, support small, local, organic farmers.

They work hard to provide food that's free from synthetic pesticides, fertilizers, and herbicides. Many of them also often have insect habitats present on their farms.

Your purchase power can extend to flowers too. Buy local, organically grown cut flowers instead of flowers grown by farms halfway around the world whose pesticide use is unknown. Buy products from companies that give a percentage of their profits to organizations that support wildlife or those who have organic gardens, green roofs, or wild spaces in their office parks and business centers. Be a smart, insect-friendly shopper whenever possible.

If you have money to donate, give money. Support your local conservation efforts or give to some of the larger conservation groups that support insects. Do you have *a lot* of money? Fund someone's research. Reach out to one of the experts in this book and tell them you'd like to donate money for their field research. They might fall over from shock, but wouldn't that be something?

GET INVOLVED

Don't have a green thumb? Feeling low on cash? That's okay. You can give the gift of your time. Help fight the fight for insects by becoming a citizen scientist. Remember me telling you in Chapter Two about that app iNaturalist? Have you installed it on your phone yet? Please do! Then try to remember to record and upload images of any and all insects you encounter. By doing so, you're creating a record for researchers to follow. You're giving them more data points so they know what's what and where that is.

Do you want to do more than just log stuff on your phone? Put yourself out there and join a butterfly count. Track bumble bees or dragonflies in North America. Count overwintering monarch butterflies in California. There are a lot of scientists who need your help. Here are a number of ongoing citizen scientist projects that you can get involved with by insect type:

MOTHS

Join John Pickering, an ecologist at the University of Georgia, Athens. He's the founder of Discover Life, an organization that promotes science education through public participation in research. The purpose of his project is to understand how ecological factors affect moths. He and Discover Life interns have been collecting data on moths in Athens, Georgia, nightly for years, amassing more than 74,000 digital photographs of moths. Of those, 71,000 have been identified to the species level, yielding a total of 925 different moth species. He's looking for help in three ways:

1. Help analyze moth phenology data, correlate with weather data, and find patterns.
2. Take photographs of moths at your porch light and upload them to Discover Life.
3. Help identify moths that you and others have uploaded so the photographs become data.

Get involved at discoverlife.org/moth.

DRAGONFLIES

Don't miss the Odonata Central website (odonatacentral.org), which collects and verifies photographic records of dragonflies and damselflies across the western hemisphere. This is a sister site with the Migratory Dragonfly Partnership, and they share data (migratorydragonflypartnership.org/index/welcome).

Entomologist Chris Goforth is looking for help with dragonflies too. She runs the citizen science project called the Dragonfly Swarm Project. She's especially interested in dragonfly swarm behaviors she has observed: static feeding swarms (dragonflies congregate in a small area to eat) and migratory swarms (massive numbers of dragonflies migrate to overwintering sites).

Find her at thedragonflywoman.com/dsp/info.

FIREFLIES

The Massachusetts Audubon Society teamed up with researchers from Tufts University to track the fate of fireflies. With your help, they hope to learn about the geographic distribution of fireflies and which environmental factors impact their abundance. All you need to do is spend at least 10 minutes once a week during firefly season observing fireflies in one location (like your backyard). All firefly sightings—or lack thereof—are valuable.

Join the project at massaudubon.org.

ANTS

Scientists want to know which kinds of ants live in urban settings, and they need your help to do it. Be a part of an international project to study ants by creating an ant picnic in your area, then reporting and uploading your results.

Start at studentsdiscover.org/lesson/ant-picnic.

LADYBUGS

Are you in the US? Entomologists at Cornell University need you to look for ladybugs. Photograph ladybugs where you find them and upload the pictures to help the scientists have detailed information on species.

Get started at lostladybug.org/participate.php.

Are you in the UK? Be a part of the UK Ladybird Survey by visiting coleoptera.org.uk/coccinellidae/home.

This website will give lots of information to help you find and identify species, then ask you to upload your observations. These researchers are creating a record of all species of ladybird found within the UK because the invasion of the harlequin ladybird (*Harmonia axyridis*) is threatening native ladybirds and other species.

POLLINATORS

The Great Sunflower Project encourages people from all over the US to collect data on pollinators in their yards, gardens, schools, and parks. By collecting visitation rates of pollinators to all plants (especially sunflowers) since 2008, this project is establishing baseline information on pollination services for the entire country and critical resources for pollinators while also identifying areas of conservation concern.

Join the project at greatsunflower.org.

BEES

Learn about several citizen scientist opportunities regarding bees at the Xerces Society site (xerces.org/community-science). You can join the Bumble Bee Watch to track and conserve bumble bees throughout the US and Canada. You can contribute to the

Nebraska Bumble Bee Atlas, a statewide monitoring effort to track and conserve native bumble bees (nebraskabumblebeeatlas.org). Or you can gather data on bumble bees throughout Oregon, Washington, and Idaho as part of the Pacific Northwest Bumble Bee Atlas (pnwbumblebeeatlas.org). BeeSpotter collects information on honey bees and bumble bees in Illinois—or be part of their annual BeeBlitz (beespotter.org). The Vermont Bumble Bee Atlas is looking to document the relative abundance and distribution of bumble bees as well as the Eastern carpenter bee (*Xylocopa virginica*) across Vermont. The survey will make essential data available to landowners, land-use planners, policy-makers, municipalities, and other individuals or organizations making conservation and management decisions.

Participate at val.vtecostudies.org/projects/vtbees.

Through the University of Florida's Native Buzz community science campaign, scientists are working with the community to learn more about the nesting preferences, diversity, and distribution of native solitary bees and wasps.

Find more at entnemdept.ufl.edu/ellis/nativebuzz/about.aspx.

BUTTERFLIES

Use Xerces Society to find opportunities to help scientists study and count butterflies (xerces.org/community-science). If you're interested specifically in monarchs, check out Save Our Monarchs at saveourmonarchs.org.

There are so many butterfly projects waiting for you, including:

- Journey North: journeynorth.org/monarchs
- eButterfly: e-butterfly.org

- Monarch Watch: monarchwatch.org
- Southwest Monarch Study: swmonarchs.org
- North American Butterfly Monitoring Network: thebutterflynetwork.org

There are ongoing insect citizen scientist projects going on literally all over the world. Scientists really need our help. Search for your favorite insect and see if you can find a study to help with. You'd be surprised what's out there. There's a Cricket Crawl in my own backyard (discoverlife.org/cricket/DC).

To my abject horror, there was even a citizen scientist project concerning cave crickets. That's right. Someone from the Rob Dunn Lab at North Carolina State University wanted us to actually report any sightings of cave crickets.[5] The data collected was used to offer insights into the "geographical distribution of camel crickets as a presence in homes."

They generated a map to show the distribution, but I cannot bring myself to look at it. The reason I mention this at all is because it is the perfect case study to illustrate how scientists need more data in order to draw accurate conclusions. This particular study concluded that cave crickets are—and I am quoting—"the most harmless, innocent creatures in the world," which so clearly cannot possibly be the truth! I think maybe these folks just did not receive enough data to tell them exactly how *repellent* and terrifying the cave cricket *really* is.

What have I left out? Oh man. Probably a lot of things. I know I've barely scratched the surface, but I hope you see that there's a

---

[5] This is actually a wonderful place. Go here to see more: robdunnlab.com/public -science.

huge role for you here as a citizen scientist. By helping researchers make observations and create data sets, you're filling an important gap in our knowledge.

FINAL THOUGHTS

I'm not an entomologist. I'm just a regular person like you. I've tried really hard to tell you everything I know about insects. I realize it's a lot. I told you stuff. Grossed you out. Filled your head. Sorry.

The one thing I hope I didn't do was make you feel bad about our planet. Yes, there are problems. Yes, insects are at risk. But we aren't powerless. There are little things we can do and bigger things we can do and, if we're really into helping, whole movements we can affect to protect insects.

We can start small. No shame in that. You read this book, didn't you? That's small. Maybe try one more thing—something I suggested or something I haven't thought up yet—to help insects. Then tell me about it, okay? It's important that we support each other. That way we can celebrate our victories and prop each other up when we aren't feeling too good about things. That might happen from time to time, you know. We will feel overwhelmed. We will feel scared or sad about the future. Of course we will. When that happens, reach out.

Above all, I want you to always think about bees. A bee is a tiny thing, right? A worker bee is going to spend her entire six-week life span working for the good of the hive. During her time on this mortal coil, she will produce about one-twelfth of a teaspoon of honey—barely enough for you to taste. The smallest of things. She's *one* bee. Yet she and her sister bees are working together always. For her hive to produce 1 *pound* of honey, they will visit *two million flowers*. One trip at a time.

That's our path too. You, me, and anyone else we can get to help us. Each of us has to Be the Bee. We have to work steady and work hard and contribute in whatever ways we can to help Mother Earth until she feels more in balance. Sure, it's daunting. There's a lot of work out there to do. But I feel better knowing that you're with me. And together, I think we can tackle this.

SUGGESTED READING LIST

I STARTED TO type up a reading list for you, but it was embarrassingly long. I consulted hundreds of books and websites and talked to countless experts to create my understanding of insects. It's too much. I didn't want to overwhelm you. So I decided—perhaps for the first time in my natural life—to show some restraint. If I could only ask you to read a handful of insect books, these are the ones I would share with you. Each is remarkable in its own way.

Although my list is short, please don't stop here. Read all the books and websites and articles you can. There's so much wonderful information out there written by people who work really hard to help us understand our world.

Bass, Bill, and Jon Jefferson. *Death's Acre: Inside the Legendary Forensic Lab the Body Farm Where the Dead Do Tell Tales.* New York: Berkley Books, 2003.

Bingham, Caroline. *Buzz!* London: Dorling Kindersley Limited, 2007.

Brackney, Susan. *Plan Bee: Everything You Ever Wanted to Know About the Hardest-Working Creatures on the Planet.* New York: Penguin Group, 2009.

Burnie, David. *E.Explore: Insect.* London: Dorling Kindersley Limited, 2005.

Eierman, Kim. *The Pollinator Victory Garden: Win the War on Pollinator Decline with Ecological Gardening.* Beverly, MA: Quarry Books, 2020.

Erzinçlioğlu, Zakaria. *Maggots, Murder, and Men: Memories and Reflections of a Forensic Entomologist.* New York: St. Martin's Press, 2000.

Goff, M. Lee. *A Fly for the Prosecution: How Insect Evidence Helps Solve Crimes.* Cambridge, MA: Harvard University Press, 2000.

Kearney, Hilary. *QueenSpotting: Meet the Remarkable Queen Bee and Discover the Drama at the Heart of the Hive*. North Adams, MA: Storey Publishing, 2019.

Lockwood, Jeffrey A. *The Infested Mind: Why Humans Fear, Loathe, and Love Insects*. New York: Oxford University Press, 2013.

Radia, Shami, and Neil Whippey. *Eat Grub: The Ultimate Insect Cookbook*. London: Francis Lincoln Limited, 2016.

Schmidt, Justin O. *The Sting of the Wild*. Baltimore: Johns Hopkins University Press, 2016.

Tallamy, Douglas W. *Bringing Nature Home: How You Can Sustain Wildlife with Native Plants*. Portland, OR: Oregon Timber Press, 2007.

Taylor, Marianne. *How Insects Work: An Illustrated Guide to the Wonders of Form and Function—from Antennae to Wings*. New York: The Experiment, 2020.

van Huis, Arnold, Henk van Gurp, and Marcel Dicke. *The Insect Cookbook: Food for a Sustainable Planet*. New York: Columbia University Press, 2014.

SELECTED WORKS CITED

I HAD A wonderful time doing the research for this book. But when I finished writing, I discovered a slight problem. I had so many sources, citations, and interview notes, the Works Cited section ran longer than the book itself! I'm confident that you can find a lot of this information on your own, so my editor and I decided not to print the list in the back of the book but to include the list on my website. You can find the sources here: brennamaloneydc.com

If you want to follow in my footsteps, this will give you a good road map to do so. You'll see a lot of fairly serious scholarly work in this list. Don't be intimidated. Most of these studies have an abstract at the beginning that will help you understand the work. It's pretty neat to see how scientists conduct their experiments and make discoveries. I'll admit that I also included a few things that I just thought were fun or interesting or too weird not to share with you. So those are in there too. If you're wondering where a specific piece of information came from and you can't find it on your own, please ask me. I'd love to hear from you, and I can point you in the right direction. You can message me here: buzzkillinsect@gmail.com.

ACKNOWLEDGMENTS

WOW. SO MANY people to thank.

When you start a project like this, it takes a long time to get up to speed—to figure out who to talk to and which questions to ask. It can be daunting. Lucky for me, I met some incredible experts along the way who were generous with their time and expertise. I would like to thank:

Dr. Guillermo J. Amador, Wageningen University and Research; Dr. Callin Switzer, research scientist at Amazon Web Services; the ever-patient, truly delightful Dr. Richard E. Walton, University College London; Dr. Wendy Lu McGill, president and CEO of Rocky Mountain Micro Ranch; Rob Allen, founder, president, and head instructor of Sigma 3 Survival School; Michael Szesze, owner and operator of the Carnivorous Plant Nursery; Dr. Steven R. Kutcher, artist and entomologist; Dr. Arthur Shapiro, entomologist, University of California, Davis; Dr. Ben Price, senior curator in charge of small orders, Insect Division, Natural History Museum, London; Dr. Nicholas Carlile, principal investigator, ecologist for the state of New South Wales Office of Environment and Heritage; Dr. Phil Mulder, professor and department head of the Department of Entomology and Plant Pathology at Oklahoma State University; Dylan Coster, math teacher, Woodrow Wilson High School, Washington, DC; Paula Goldberg, City Wildlife board member and former executive director, Washington, DC;

Toni Burnham, DC Bee Alliance; and Kathy Lally, *The Washington Post*.

There's a whole group of people in the publishing world who not only have to believe in you and your idea but also must work tirelessly to edit, fact-check, publish, market, promote, and sell it on your behalf. I'm very grateful to my agent, Helen Adams, and my editor, Laura Godwin. Thank you for not asking me to be something else or to sound like someone else. Thanks to the entire team at Henry Holt Books for Young Readers.

It is a wonderful honor to have the brilliant work of Dave Mottram in this book. How could anyone not love insects after seeing his delightful illustrations? Thank you, Dave, for capturing them in a way that transcends my meager words.

Writing can be a lonely endeavor because you have to spend a lot of time in your own head, and you spend an even longer amount of time hunched over a laptop, typing, typing, typing. It helps to have a lot of support.

For those of you who bravely read bits and pieces of the manuscript as it was being written, I am in your debt. Especially my mom, who read every word. Thanks also to Abrar Mohamed. I'm not sure I can forgive you for opening my world up to the concept of cockroach milking (*eeew gross*), but your enthusiasm for my work kept me writing. To my tribe at Glen Echo—Jan Wickham, Cathy Orme, and Katy Orme. Words fail me. When I struggle in life, you carry me. When I fly, you soar with me.

Thank you to Fran Jacobs and her colleagues at CSA Creative for some early fact-checking work on my behalf. Such a comfort to have you guys watching my back. Fran, your friendship and support meant so much to me during the creation of this book.

And to my family:

Liam and Devin: I know you put up with a lot. I brought insects home with me. I asked you to eat some of them. I talked about them nonstop for, quite literally, *years*. I'm sorry about that. I hope I didn't scar you two with all my shenanigans. When you are middle-aged men and in therapy, you can tell the doctor that it was me who told you that unfortunate business about bee penises. Still and all, it meant the world to me to share this work with you.

Chuck: Thanks for putting up with the deer head incident. And the cockroaches. And a whole lot of non-insect-related things over the last 20 years. For good or for ill, you've always just let me be me, and that's more than anyone could ask for.

Mom and Dad: You must inwardly cringe every time you hear me say, *I had this idea* . . . And yet you are unflinchingly with me every step of the way. Your faith in me means everything.

And to Steve: This is all *your* fault, you know. Never in my wildest would I have attempted such a big project on my own. Somehow you just made me think I could do it.

INDEX